HIGH VACUUM SERIES

Edited by L. HOLLAND F.Inst.P.

HIGH VACUUM TECHNIQUE

HIGH VACUUM TECHNIQUE

Theory, Practice and Properties of Materials

by

J. YARWOOD

M.Sc., F.Inst.P.

Professor of Physics,
The Polytechnic of Central London

FOURTH EDITION

(*Completely revised*)

CHAPMAN AND HALL

and

SCIENCE PAPERBACKS

First published 1943
Second edition 1945
Third edition 1955
Fourth edition 1967

First published as a Science Paperback 1975 by
Chapman and Hall Ltd
11 New Fetter Lane, London EC4P 4EE

Made and printed in Great Britain by
William Clowes and Sons, Limited
London, Colchester and Beccles

ISBN 0 412 02520 5 (*cased edition*)
ISBN 0 412 21190 4 (*Science Paperback edition*)

Distributed in the U.S.A. by
Halsted Press, a Division of
John Wiley & Sons, Inc.,
New York

PREFACE TO THE FOURTH EDITION

This book appeared in successive editions in 1943, 1945 and 1955. Each time it was enlarged but the text was essentially based on that first written in 1943. Within twenty-three years the subject has developed so much that producing this present edition has had to involve rewriting the text completely.

The chief purpose of this book is to introduce the theory and practice of vacuum to students of science and technology at undergraduate and immediate post-graduate levels. A knowledge of elementary physics and introductory kinetic theory has been assumed, but most of the text could readily be understood by a reader without this background. The emphasis is on the apparatus and methods more likely to be encountered in the college rather than the industrial laboratory, in that applications of vacuum have been largely ignored in favour of dealing with fundamental principles. However, the importance of incorporating into educational practice the ideas involved in recent industrial methods has been considered throughout the text.

The final chapter gives data on miscellaneous materials, processes and devices frequently encountered in vacuum technique. This is of necessity far from complete; it is included because of the importance of this kind of information to the college or industrial research worker.

Sincere thanks are due to Mrs Angela Williams for typing the manuscript and to Mr Ronald Dewar who drew most of the diagrams.

J. YARWOOD

CONTENTS

NOTE

The unit of pressure used throughout this book is the torr. The torr is equal to the millimetre of mercury (mm Hg) to within 1 part in 7×10^6.

1 standard atmosphere = 760 torr
1 micron (μm) of mercury = 10^{-3} torr
1 microbar = 1 dyne per sq cm = $0{\cdot}75 \times 10^{-3}$ torr

CHAPTER ONE

THE PRODUCTION OF VACUA

1.1. *Introduction*

The **standard atmospheric pressure** (atm) is 760 millimetre of mercury. In this text, the term torr will be used as the unit of pressure, equal to 1 millimetre of mercury (mm Hg). A perfect or absolute vacuum implies the unrealizable state of space entirely devoid of matter. For practical purposes, the term 'vacuum' may be used to denote roughly gas pressures below the standard atmospheric pressure, i.e. below 760 torr. To be more specific, Table 1.1 shows the accepted terminology in denoting degrees of vacuum and the pressure ranges concerned.

TABLE 1.1

Degrees of vacuum and pressure ranges

Degree or quality of vacuum	*Pressure range in torr*
Coarse or rough vacuum	760 to 1
Medium vacuum	1 to 10^{-3}
High vacuum	10^{-3} to 10^{-8}
Ultra-high vacuum	$<10^{-8}$

$$10^{-3} \text{ torr} = \text{torr}/1{,}000 = \text{millitorr (mT) or micron } (\mu\text{m})$$
$$10^{-6} \text{ torr} = \text{torr}/1{,}000{,}000 = \text{microtorr } (\mu\text{T})$$

Pressure is force per unit area. In the centimetre-gram-second (c.g.s.) system of units, the unit of pressure is consequently the dyne per sq cm, i.e. dyne/cm^2 or dyne cm^{-2}. This unit is also known as the **microbar.**

$$1 \text{ torr} = 1 \text{ mm Hg} = 0{\cdot}1\times13{\cdot}6\times981 \text{ dyne cm}^{-2} = 1333 \text{ microbar}$$

$$10^{-3} \text{ torr} = 1 \text{ millitorr (mT)} = 1 \text{ micron of mercury} = 1{\cdot}333 \text{ microbar}$$

So,

$$1 \text{ microbar} = 1 \text{ dyne cm}^{-2} = 0{\cdot}75\times10^{-3} \text{ torr}$$

In the metre-kilogram-second (m.k.s.) system of units, the unit of pressure is the newton per sq metre, i.e. newton/m^2 or newton m^{-2}.

$$1 \text{ newton} = 10^5 \text{ dyne, and } 1 \text{ m}^2 = 10^4 \text{ cm}^2$$

$$\therefore \quad 1 \text{ newton m}^{-2} = 10^5/10^4 = 10 \text{ dyne cm}^{-2} = 10 \text{ microbar}$$
$$= 0{\cdot}75 \times 10^{-2} \text{ torr}$$

$$\therefore \quad 1 \text{ torr} = 133 \text{ newton m}^{-2}$$

In France, the newton per sq metre (newton m^{-2}) is frequently called the **pascal**. Thus,

$$1 \text{ torr} = 133 \text{ pascal}$$

and

$$1 \text{ pascal} = 7{\cdot}5 \text{ millitorr}$$

When gas is removed from a vessel by some means of pumping, the gas pressure in the vessel falls, the number of molecules per unit volume in the vessel decreases, and the average separation between the molecules increases. For an ideal gas, the pressure, p, and the number of molecules per unit volume, n, are related by the equation

$$p = n\text{k}T \tag{1.1}$$

where T is the absolute temperature and k is Boltzmann's constant. At a given temperature, therefore, the number of molecules per unit volume is directly proportional to the pressure.

In relation to the average separation between the molecules, a most useful quantity is the **mean free path** (m.f.p.). This is defined as the average distance a molecule travels, at the prevailing pressure, between successive collisions with other molecules. Let d be the diameter of a molecule. If two molecules approach to within a distance d between their centres, we may say they have collided. This is equivalent to denoting the target area of a molecule as πd^2. If there are n molecules per unit volume, and the number within a volume of cross-sectional area S and length L is nSL, their total target area is $\pi d^2 nSL$. When $\pi d^2 nSL = S$, the target formed by the molecules fills the whole area S, so that no molecule can traverse a distance L without making a collision. The value of L which makes $\pi d^2 nSL = S$ is

$$L = \frac{1}{\pi d^2 n}$$

which therefore should give the m.f.p., L.

However, in this simple deduction, it is assumed that a molecule of a given velocity is traversing the gas and the other molecules are at rest. Clausius made a first approximation to the real case and improved the theory by assuming that all the molecules move with the same velocity. He obtained the result

$$L = \frac{1}{\frac{4}{3}\pi d^2 n}$$

However, there is a Maxwellian distribution of velocities amongst the molecules, which, when taken into account, modifies this result to

$$L = \frac{1}{\sqrt{2}\,\pi d^2 n} \tag{1.2}$$

where L is the m.f.p.

At a given temperature T, equation (1.1) shows that n is proportional to p. Therefore, for a given type of gas of constant molecular diameter d,

$$L = k/p \tag{1.3}$$

where k is a constant. So the m.f.p. is inversely proportional to the pressure at constant temperature for a given gas.

For nitrogen (and hence approximately for air), a useful practical form of equation (1.3) is

$$L = 5/p \text{ cm} \tag{1.4}$$

where p is the pressure in millitorr.

Thus, for air at 1 torr, L is 5×10^{-3} cm; at 10^{-3} torr, it is 5 cm; and at 10^{-6} torr, it is 5,000 cm.

The m.f.p. is a most valuable concept because it gives a measure of how readily, or otherwise, particles will travel through the gas. In the passage of electrons through a gas, the m.f.p. for an electron will be longer than for a molecule, because the electron is vanishingly small compared with the molecule. The m.f.p., L_e, for an electron passing through a gas is given by

$$L_e = 4\sqrt{2}\,L = 5{\cdot}66L \tag{1.5}$$

L being the m.f.p. of the molecules. It follows from equation (1.5) that for electrons in attenuated air

$$L_e = \frac{5 \times 5{\cdot}66}{p} \text{ cm} = \frac{28{\cdot}3}{p} \text{ cm}$$

In an electron tube in which the residual gas pressure is, say, 10^{-5}

torr, assuming this gas is nitrogen (which, in fact, is not usually the case), the m.f.p. of the electrons will be

$$L_e = \frac{28{\cdot}3}{10^{-2}} = 2\,830 \text{ cm}$$

Consequently, the electrons can travel between electrodes, say 5 cm apart, with a small probability of colliding with gas molecules, and hence there is little ionization of the gas.

It is, however, stressed that L and L_e are mean values. In the derivation of L as a mean value leading to equation (1.2), the target area, πd^2, of any one molecule is multiplied by n, the number of molecules per unit volume, to give $\pi d^2 n$ as the cross-sectional target area per unit volume. This assumes a uniform distribution of molecules within a given volume. In reality, because of molecular chaos, the free path of individual molecules varies from zero to infinity, and it can be shown that

$$n_m = n\mathrm{e}^{-l/L} \tag{1.6}$$

where n_m is the number of those molecules in unit volume (containing n molecules) able to traverse a distance l in the gas without collision; L is the m.f.p., and e is the exponential base 2·71828. For example, in the case of electrons in nitrogen at a pressure of 10^{-5} torr, the fraction f of the number of electrons able to travel a distance of, say, 2 cm unobstructed by gas molecules is

$$f = n_m/n = \mathrm{e}^{-2/2830} = \mathrm{e}^{-0{\cdot}0007} = 0{\cdot}9993$$

This is almost all the electrons; there is, nevertheless, a fraction, 0·0007 (1 – 0·9993), which does make collisions, and this may well be significant.

Under static pressure conditions, the molecules of a permanent gas have random kinetic motion in all directions, with an average velocity $\bar{v}$ given by

$$\bar{v} = \sqrt{\left(\frac{8\mathrm{k}T}{\pi m}\right)} = \sqrt{\left(\frac{8\mathrm{R}T}{\pi M}\right)} \tag{1.7}$$

where k is Boltzmann's constant, R is the universal gas constant, T is the absolute temperature, m is the mass of a molecule, and M is the molecular weight. For nitrogen at $T = 300°$K, substituting R = $8{\cdot}317 \times 10^7$ erg per degC per mole, and $M = 28$,

$$\bar{v} = \sqrt{\left(\frac{8 \times 8{\cdot}317 \times 10^7 \times 300}{\pi \times 28}\right)} = 4{\cdot}8 \times 10^4 \text{ cm sec}^{-1}$$

The rate at which gas molecules impinge on unit area of a boundary in a gas is often important in vacuum practice, because the molecules may stick to or condense on the boundary; further, it will decide the rate at which the molecules can traverse an aperture in a solid boundary (section 3.2). A result from Knudsen's work in the kinetic theory of gases is that

$$N = \tfrac{1}{4}n\bar{v} \tag{1.8}$$

where n is the number of molecules per unit volume in the gas, and N is the number of these molecules which impinge on unit area of a boundary in the gas per sec. Here it is assumed that the m.f.p. of the molecules is greater than the dimensions of the gas enclosure concerned, i.e. the gas is in the region of the so-called **molecular pressures** which, generally speaking, prevails below 10^{-3} torr.

Substituting for $\bar{v}$ from equation (1.7), and for n from equation (1.1), equation (1.8) becomes

$$N = \frac{p}{4\mathrm{k}T}\sqrt{\left(\frac{8\mathrm{R}T}{\pi M}\right)}$$

Putting $\mathrm{k}=\mathrm{R}/\mathrm{N}$, where R is the gas constant per mole, and N is Avogadro's number, this equation becomes

$$N = \frac{p\mathrm{N}}{\sqrt{(2\pi M\mathrm{R}T)}} \tag{1.9}$$

Substituting $\mathrm{N}=6{\cdot}025\times10^{23}$, $\mathrm{R}=8{\cdot}317\times10^{7}$ erg per degC per mole, $T=293$°K (20°C), and $M=28$ for nitrogen,

$$N = \frac{6{\cdot}025\times10^{23}p}{\sqrt{(2\pi\times28\times8{\cdot}317\times10^{7}\times293)}}$$
$$= 2{\cdot}9\times10^{17}\,p \text{ (impacts cm}^{-2}\text{)}$$

At a pressure of, say, 10^{-6} torr, equivalent to $1{\cdot}333\times10^{-3}$ dyne per sq cm, the number of molecular collisions per sec with a boundary of area 1 sq cm is therefore $2{\cdot}9\times10^{17}\times1{\cdot}333\times10^{-3}$, i.e. $3{\cdot}86\times10^{14}$.

VACUUM PUMPS

Pumps for the production of vacua may be roughly divided into two classes: (*a*) those which pump air or gas from a vessel which can be at atmospheric pressure initially; (*b*) those which can only begin to operate below a certain pressure considerably less than atmospheric, and hence require a **fore-vacuum** (also known as a **backing vacuum**).

Within this broad classification, there is a wide variety of types, but the most commonly encountered ones within category (*a*) are the **water-jet pump**, the **oil-sealed mechanical pump**, and the **sorption pump**; and within category (*b*), the **vapour pump**, the **molecular drag pump**, the **Roots pump**, the **getter-ion pump**, and the **cryopump**. In general, in a vacuum system able to produce pressures below 10^{-3} torr, the most common arrangement is a pump of the oil-sealed mechanical type backing a vapour pump. The former discharges to the atmosphere, and the vapour pump in series discharges to a back-

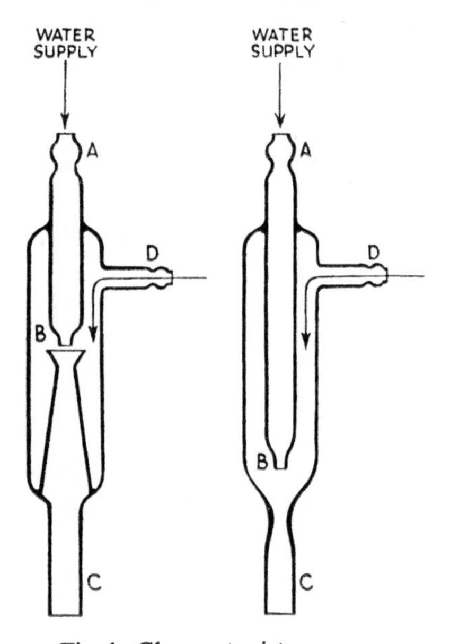

Fig. 1. Glass water-jet pumps.

ing vacuum created by the mechanical pump. A second arrangement currently of increasing popularity is a sorption pump as a backing stage to a getter-ion pump.

The **speed of a pump** is measured at a given pressure and equals the volume of gas pumped from a vessel in unit time at that pressure, where the volume is measured at the pressure (see also section 4.1).

1.2. *The Water-jet Pump*

Water supply from a fast-running tap is fed through thick-walled rubber tubing into the connection at A (Fig. 1). This water-stream emerges at high speed from the converging jet at B. This jet is surrounded by a cone to prevent splashing and also to guide the water-stream down to waste at C. A side-tube, D, is connected (usually by rubber pressure tubing) to the vessel to be evacuated. Molecules of nitrogen and oxygen in the region of B are trapped by the high-speed jet and forced into the atmosphere; these molecules are replaced by further ones from the vessel undergoing pumping, and so on. By this means, a vacuum is created which is limited chiefly by the saturated vapour pressure of water, which is 17·5 torr at 20°C. The use of a drying agent (silica-gel or phosphorus pentoxide) placed in a suitable container within the tube connecting the pump to the vessel enables lower pressures to be obtained, but with a glass pump the limit is about 5 torr. The pumping speed is of the order of 1·2 litre per min at a pressure of 10 torr for a glass pump, but steel, and also plastic, water-jet pumps of superior performance and robustness are commercially available. For example, Edwards High Vacuum Ltd. market water-jet pumps having stainless steel jets and either a metal or plastic body, which provide a pumping speed of 3·5 litre per min, when the water pressure supplied is 20 lb per sq inch. One of these pumps will evacuate a vessel of 5 litre volume to an ultimate pressure of about 12 torr in 12 min.

1.3. *Oil-sealed Mechanical Pumps*

There are several designs, but they have the common property that an electric-motor-driven rotating cylinder or plunger sweeps the gas from a region A, and discharges it to a region B which is usually at atmospheric pressure. A gas-tight seal is continuously effected between A and B. The first practical vacuum pump of this type was designed by W. Gaede [1] and marketed by Leybold's Nachfolger in Germany in 1910. This was a **rotary vane pump**; the principle of operation is illustrated by Fig. 2(*a*). A modern version, typical of the Edwards High Vacuum Ltd. 'Speedivac' series of pumps, is shown in Fig. 2(*b*).

The rotor, which consists of a solid steel cylinder (Fig. 2*a*), rotates on an axle inside a cylindrical steel stator. The drive to the central rotor axle is by means of a pulley-coupled electric motor. A and B are the intake and discharge ports respectively, i.e. the container being evacuated is joined to A and the air is sucked from it and discharged to the atmosphere at B. Over part of the area about the top-most point of the stator, there is a minimum clearance between the rotor and stator surfaces of about 0·002 inch. The rotor is driven about its

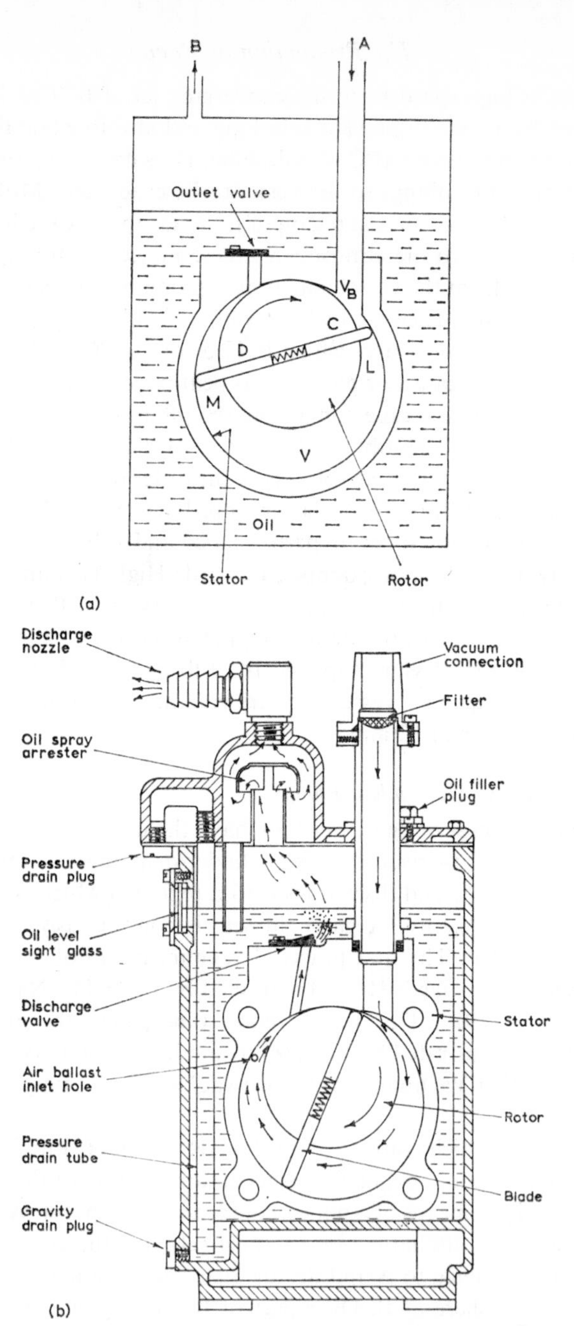

Fig. 2. (*a*) Principle of operation of an oil-sealed mechanical pump of the rotary vane type. (*b*) A 'Speedivac' rotary pump.

axle in the direction indicated by the arrow. Two vanes or blades, C and D, are inserted in diametrical slots in the rotor. The outside edges of these vanes are forced against the inside wall of the stator by means of central springs. The surfaces of the rotor and the stator are ground to a high degree of precision. The assembly is immersed in oil in the pump casing; this oil acts as a seal and as a lubricant.

As the rotor rotates, the leading edge L of one of the vanes, C, passes the junction between the stator and the intake port. Subsequently during the rotation, the volume V_B behind this vane expands, so the pressure at the intake decreases. Whereas the crescent-shaped volume V in front of this vane begins to decrease *after* the leading edge M of the opposite vane, D, passes the discharge outlet, i.e. during a half revolution the gas in volume V is decreased. When the leading edge L of vane C in turn reaches the junction to the discharge outlet, the gas will be compressed to a minimum residual volume. At this volume, its pressure is high enough to force open the one-way valve in the discharge outlet against the pressure created by the small retarding force of the valve, the small head of oil above the outlet and the pressure of the atmosphere. Hence, in each revolution, the outlet valve opens twice: once when leading edge L passes it and a second time on the passage of M. Each time, a volume of gas, defined by the maximum volume V enclosed to one side of the vanes between the rotor and the stator, and at the pressure at the intake, is compressed and swept out into the atmosphere. If the number of revolutions per minute is n, S_D, the **free air displacement** of the pump, which is its speed at atmospheric pressure, is given by

$$S_D = 2nV \text{ litre min}^{-1}$$

where V is in litre.

The outlet valve to the discharge port may be of various designs: in small pumps a simple steel blade or a neoprene flap is common; larger capacity pumps frequently have a spring-loaded ball outlet valve.

As the pumping speed is quoted at the pressure prevailing at the inlet, it would be expected that this speed were constant irrespective of the pressure, because V is constant. However, as the lower pressures are attained, there is in effect a leak of gas into the pump. This is due to: the small but necessary minimum clearance between the rotor and stator; the finite though very small minimum swept volume between the rotor and stator; the vapour pressure of the oil; and the fact that this oil tends to decompose into gases, especially at localized higher temperature spots in the pump. These factors will result in the pumping speed falling below the free air displacement at pressures below about 10 torr, eventually to become zero at the ultimate pressure (the lowest pressure the pump is able to attain), which is 10^{-2} torr approximately.

A lower ultimate pressure of approximately 10^{-4} torr can be achieved by using two such pumps in series; the discharge outlet of the first being connected to the intake of the second and the discharge of the second being to the atmosphere. The typical characteristic of pumping speed against pressure for a single-stage pump is shown in Fig. 3. Superimposed on the same axes is that of a two-stage model of

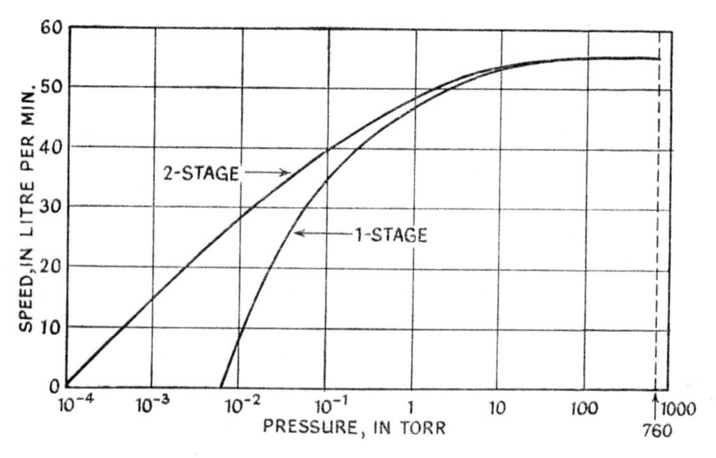

Fig. 3. Pumping speed against pressure characteristics of a single-stage and a two-stage oil-sealed rotary pump.

the same free air displacement but considerably lower ultimate. Two-stage models are commercially available where the interconnected stages have a common axle driven from the same motor and are immersed in oil in a single housing (Fig. 4).

Among the several alternative types of mechanical rotary pump, one frequently encountered is a model originated by the Central Scientific Co. in America. Their 'Cenco-Hyvac' design is illustrated in principle by Fig. 5. It utilizes a single spring-loaded vane bearing on the surface of an eccentrically-mounted rotor inside the stator; its mode of operation can be easily deduced as it is basically similar to the two-vane type already described. A second alternative, the

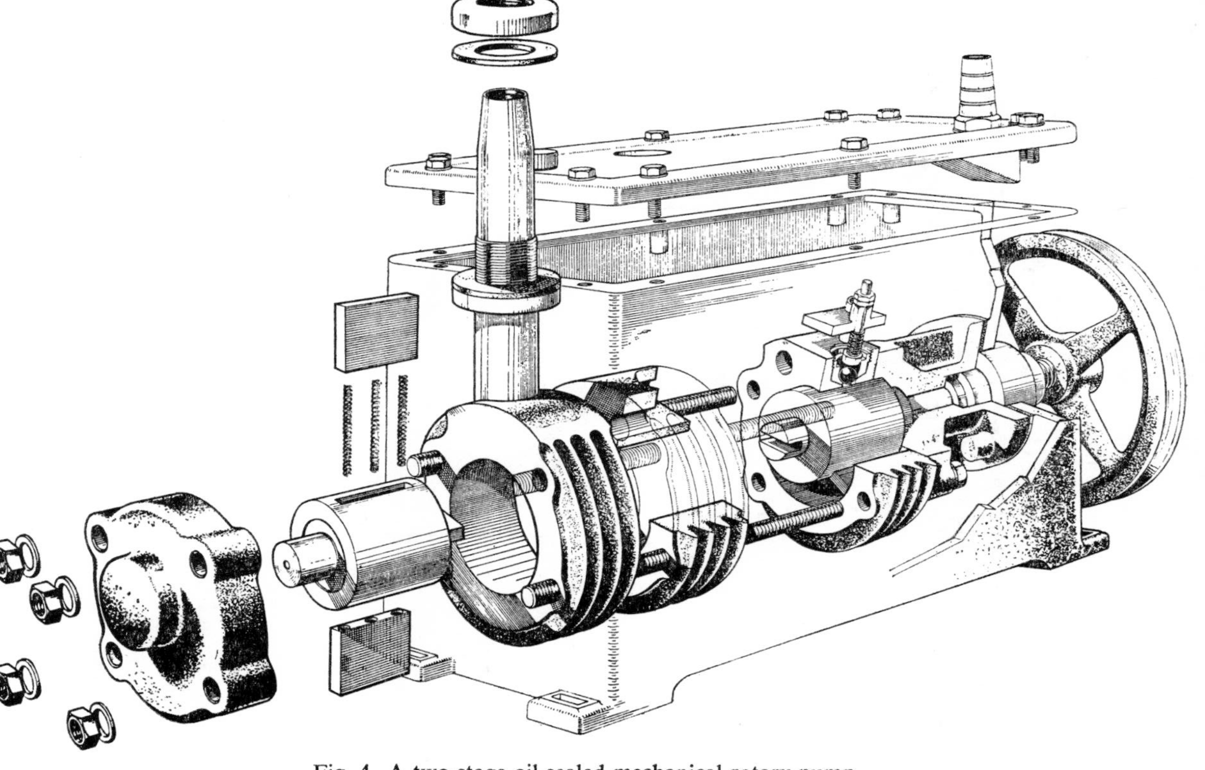

Fig. 4. A two-stage oil-sealed mechanical rotary pump.

rotating plunger pump is, however, considerably different in operation and is a popular model in the larger sizes. Indeed, the rotary vane pump is not generally built in sizes exceeding a free air displacement of 3,000 litre per min, whereas rotating plunger pumps of considerably greater displacements are made.

This rotating plunger pump (Fig. 6) was introduced in America by the Kinney Manufacturing Co. Instead of making use of vanes, it has a tube, F, of rectangular cross-section which is a sliding fit in a small auxiliary cylinder and connects the intake port to the rotor or plunger. This plunger is mounted eccentrically about the motor-driven,

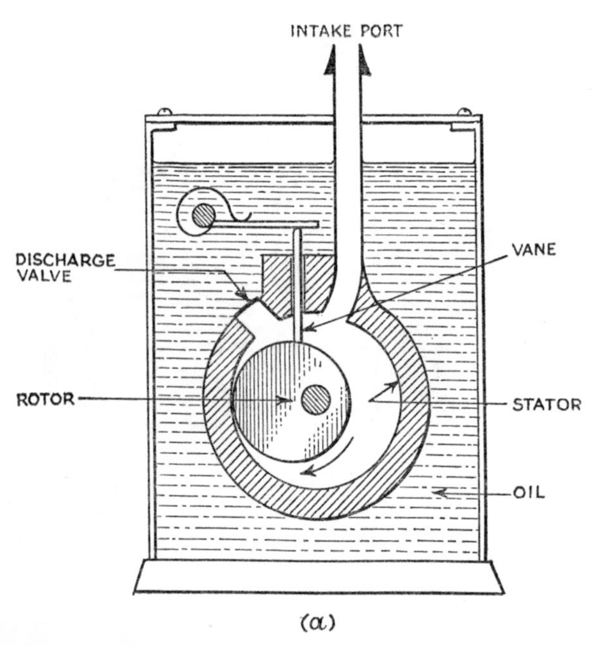

Fig. 5. Principle of the Cenco-Hyvac mechanical rotary pump.

revolving axle E and is in two parts: the inner drum C rotates about the axle E, but the cylindrical shell D is a sliding fit on C and, because it is attached rigidly to the inlet sliding tube F, will not rotate with C but undergoes a 'rocking' motion, whereby the point G, at which there is close contact between the plunger and stator, sweeps round the inner wall of the stator.

As the plunger moves in the direction of the arrow, it rapidly creates extra space at A into which is admitted some of the gas from the container connected to the intake port. Simultaneously, compression of the gas previously trapped in volume B is taking place. When the plunger has almost reached its highest point, it expels all air or gas and surplus sealing oil through the feather-type outlet valve and nozzle H into the oil separator tank; here the oil is retained and the air or gas is discharged into the atmosphere. With further movement of the plunger, the intake port is completely closed, and

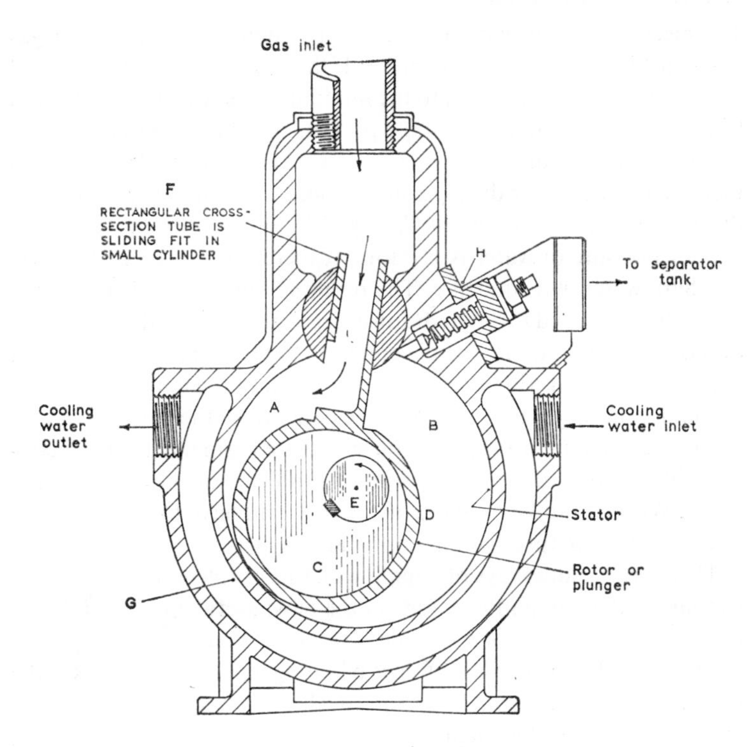

Fig. 6. The Kinney rotating plunger pump.

the air or gas admitted to the full space A is trapped and then expelled during the next revolution, and so on.

A wide range of sizes of mechanical rotary pumps is commercially available from single- or two-stage models (with a free air displacement of 20 to 30 litre per min, oil immersed, and with drive at 400 to 700 r.p.m. from an $\frac{1}{8}$ h.p. electric motor) to large installations of the Kinney and other such types (with a displacement of about 20,000 litre per min, a 40 h.p. electric motor drive, and consisting of two stages in parallel). Pumps with displacements below 200 litre per min are usually oil-immersed patterns, i.e. the rotor-stator assembly is entirely within a low vapour pressure oil in a metal box housing. Larger pumps generally have a continuously-circulated oil feed from a separate reservoir; the feed-rate being adjustable. Further, the large models are often water cooled.

1.4. *The Gas Ballast Pump*

The mechanical rotary pump operates by compressing the air or gas and expelling it into the atmosphere. If the compression ratio, defined as the ratio of the maximum to the minimum swept volume, is C; and free water is present at the intake port at, say, 20°C, so that it exerts its saturated vapour pressure of 17·5 torr, this pressure will tend to increase to 17·5C torr during compression. The pump will run warm during operation at about 60°C, at which temperature the saturated vapour pressure of water is 150 torr. If therefore C exceeds 150/17·5 (i.e. 8·6), water will condense. The resulting liquid, together with the gas, will be expelled through the discharge valve, mix with the oil, and evaporate again into the vessel connected to the intake port; the total pressure produced will be seriously increased. In the normal mechanical pump, the larger the value of C the better from the point of view of creating a vacuum, if the gases concerned are permanent; but, if condensable vapours are present, serious problems of oil deterioration result. With an intake pressure of 1 torr, C will have to exceed 760 to discharge the compressed gas to the atmosphere.

There are various ways of coping with this problem of condensable vapours at the pump intake. One of the most useful and certainly the most ingenious is due to W. Gaede [2] and was first adopted in pumps marketed by Leybold's Nachfolger of Cologne, Germany. This is the use of gas-ballast. Nowadays, almost all commercial rotary pumps are provided with this facility.

In the provision of gas-ballast on a mechanical pump, condensation of water (and other vapours) is greatly reduced by admitting air from the atmosphere to the pump via a one-way gas-ballast valve of the ball type, at a suitable time in the revolution of the rotor. The ratio by which the air-vapour mixture is compressed is thereby reduced to between 6:1 and 10:1. Then, if water vapour enters the pump intake at less than 30°C (saturated vapour pressure of water = 32 torr), whereas the pump interior is at 60°C (vapour pressure = 150 torr) or more, the water does not condense at all, but is discharged as vapour with the permanent gases through the outlet valve of the pump. This gas-ballast also helps by maintaining a rather higher pump interior temperature of about 70°C.

The now severely-limited compression ratio, *C*, means, of course, that the ultimate pressure provided by the pump is considerably higher. Indeed, for a single-stage pump, it is increased at least a hundred-fold, from about 10^{-2} to 1 torr or more. However, the air inlet via the gas-ballast valve can be reduced by simply screwing-in the valve. In many practical cases, water vapour is present chiefly in the initial stages of pumping. Therefore, full gas-ballast is applied when pumping is begun. The water vapour present is largely removed, and finally the gas-ballast valve is shut, so the full compression ratio, and hence the ultimate pressure of 10^{-2} torr, is attainable. In two-stage mechanical pumps, only the backing stage to the atmosphere is gas-ballasted. The ultimate pressure is then not so adversely affected. Typically, a two-stage pump with full gas-ballast will attain 10^{-2} torr, which is reduced to about 5×10^{-4} torr when the gas-ballast valve is closed.

To admit the ballast air to the pump (dry air is desirable, otherwise more ballast air is needed; in a laboratory where the relative humidity is high, a simple drying tube in the air inlet is good practice) at the correct time in the operating cycle of the pump rotor, the air inlet to the ball valve is adjustable by a screw (Fig. 7). Consider the position of the leading vane A. The gas-ballast inlet is open. Evacuation occurs over the part of one revolution from positions 1 to 2, because, when this vane is at 1, the space between the rotor and stator *below* the vanes is isolated from the intake port, and the space *behind* this vane A increases in volume. When the leading vane A arrives at position 2, however, the rear vane B is at position 1, so that the space *behind* the leading vane A becomes isolated from the intake port. During the part of the rotation from 2 to 3, this isolated gas is not compressed, so the ball in the gas-ballast inlet valve is down, i.e. this valve is open, because the atmospheric pressure acting on the ball exceeds the pressure of the isolated gas-vapour mixture in the pump. As rotation of A continues beyond 3, the isolated gas and vapour, and also the air admitted via the gas-ballast valve, are compressed; this forces the ball upwards, so that the gas-ballast valve is closed, and consequently the discharge valve opens to release the gas-vapour mixture to the atmosphere. It is thus ensured that the gas is not simply re-expelled through the gas-ballast inlet. The discharge valve is now opened chiefly by the pressure due to the ballast air and not the air from the vessel which has passed through the pump; hence the partial pressure of the vapour in the mixture does not attain the saturated value, so condensation does not occur.

Though condensable vapours may be satisfactorily removed from a container by gas-ballast pumps, those vapours which react chemically with the pump oil or metal parts should be avoided.

Typical pumping speed against pressure characteristics for a single-stage rotary pump with and without gas-ballast are shown in Fig.

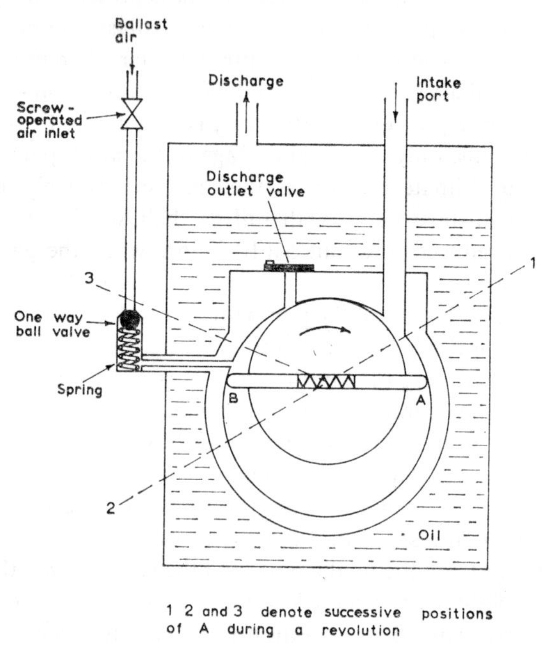

Fig. 7. Operation of gas-ballast on a rotary pump.

8(*a*); the accompanying Fig. 8(*b*) shows the advantage gained by using gas-ballast when evacuating a mild steel vessel. Even though such a container is superficially dry inside, water vapour is invariably released by the walls, and this is only removed some time after using the gas-ballast valve. Only then does the total pressure as recorded by a thermocouple gauge (section 2.4) become equal to the partial pressure of the permanent gas as measured by a McLeod gauge (section 2.3).

There is, as would be expected, a limit to the rate at which a gas-ballast pump can pump a vapour without the pump oil becoming contaminated.

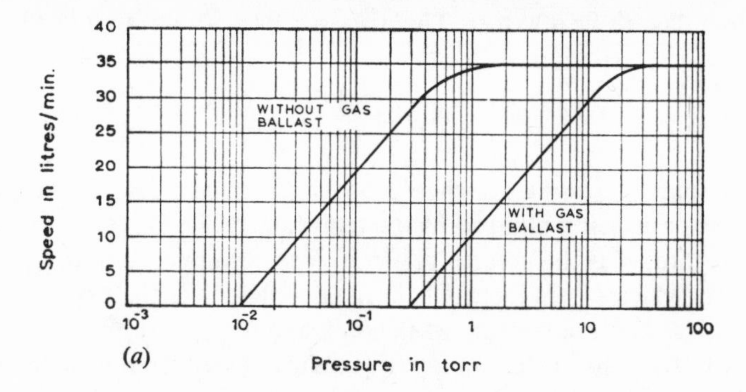

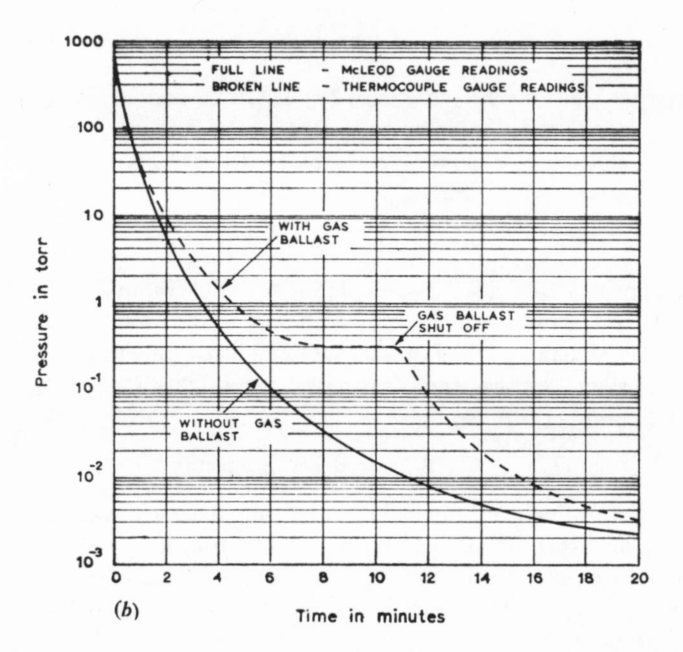

Fig. 8. (*a*) Pumping speed against pressure characteristics for a single-stage mechanical pump with and without gas-ballast. (*b*) Pressure-time curve obtained on pumping a steel vessel.

To calculate this rate (Power and Kenna [3]), it is assumed that vapour alone is being pumped, and, at the limit, the ballast air just becomes saturated with vapour at the pump temperature, T_2°K, as the discharge outlet valve begins to open. This opening of the valve will occur at a pressure, p_e, somewhat above atmospheric pressure, usually taken to be 900 torr. The mass per min of emerging vapour is then

$$\frac{V\rho p_a}{p_e - p_v}\frac{T_2}{T_1}$$

at atmospheric pressure p_a; where p_v is the saturated pressure, in torr, of the vapour being pumped at the temperature T_2; T_1 is the room temperature; V is the volume of air, in litre per min at atmospheric pressure, which enters the gas-ballast valve; and ρ is the density of this vapour in gram per litre at the temperature T_2 and pressure p_v. This will equal the maximum safe throughput Q_v of vapour in gram per min, i.e.

$$Q_v = \frac{V\rho p_a T_2}{(p_e - p_v)T_1} \qquad (1.10)$$

Typical values for T_1 and T_2 are 293°K (20°C) and 343°K (70°C) respectively. For water vapour, p_v is 234 torr at 70°C, so $(p_e - p_v)$ is 666 torr. Putting $p_a = 760$ torr, equation (1.10) becomes

$$Q_v = \frac{\rho V \times 760 \times 343}{666 \times 293} = 1{\cdot}33\rho V \text{ gram min}^{-1}$$

An approximate value for ρ, the density of water vapour, at 343°K and 234 torr, can be evaluated by assuming water vapour to be an ideal gas and making use of the fact that 22.4 litre at 760 torr and 0°C has a mass of 18 gram (the gram molecular weight). On this basis, ρ is 0·2 gram per litre, so that

$$Q_v = 1{\cdot}33 \times 0{\cdot}2\text{V} = 0{\cdot}27V \text{ gram min}^{-1}$$

The dry air flow with the ballast valve fully open is about 10% of the free air displacement of the pump. A pump with a displacement of 1,000 litre per min will therefore have a maximum air ballast flow of 100 litre per min. The maximum allowable amount of water vapour it can pump at 70°C is therefore

$$Q_v = 0{\cdot}27 \times 100 = 27 \text{ gram min}^{-1}$$

1.5. *Vapour Pumps*

In attempting solutions of the problem of creating a vacuum, the possibility exists of driving gas from one region A to another B by means of a swiftly-moving stream of vapour. Thereby, a pressure gradient may be established in the gas, so that its pressure at A, p_A, is considerably less than the pressure at B, p_B. Here, p is the partial pressure of the gas being pumped and does not include the pressure of the vapour in the stream.

Under static pressure conditions, the molecules of a permanent gas have random kinetic motion in all directions with an average velocity $\bar{v}$, given by equation (1.7), which, for nitrogen at 300°K (27°C), was shown to be about 500 metre per sec.

To drive these molecules, of average speeds of about 500 metre per sec, in a preferred direction therefore requires a high velocity vapour stream. Indeed, a suitable liquid – the **pump fluid** – must be heated to such a temperature that its vapour issues, from a nozzle or jet, as a stream in which its molecular velocities are predominantly in a given direction, from A to B, and in which these velocities exceed considerably the mean kinetic speed of the gas molecules, i.e. in the supersonic region. Furthermore, this pump fluid must be prevented as far as possible from entering the chamber connected to A, otherwise it will contribute its vapour pressure at the prevailing temperature to the low pressure created. Apart from designing the pump so that the vapour stream travels from A to B, and with as little back-streaming as possible in the opposite direction, i.e. from B to A, it is also necessary, as back-streaming cannot be completely eliminated, either to have a pump fluid with a very low vapour pressure and/or to reduce or prevent back-streaming by a baffle or trap.

The problem has been successfully solved in the widely used vapour pumps. Two main classifications of this type of pump are **vapour diffusion pumps** and **vapour ejector pumps**. There are also pumps which are combinations of these and, further, those in which both diffusion and ejector actions occur. Possible pump fluids are mercury, low vapour pressure oils, and steam. The last of these is successfully used in **steam-ejector pumps**, which have fairly widespread application in specialized branches of engineering where large displacements at pressures down to 0.1 torr or lower in a multi-stage system are needed. However, they require a steam supply at a pressure of some 60 lb per sq inch, so are outside normal laboratory practice and will not be further discussed here. There is also the possibility of a pump employing the vapour of a metal other than mercury, but the problems of heating the metal, choosing a metal of low chemical reactivity, condensing the vapour, and arranging recirculation controllably in suitable regions of the pump have so far precluded satisfactory development.

The most important vapour pumps are thus the **mercury diffusion pump**; the **oil diffusion pump**, the **mercury ejector pump**, and the **oil**

ejector pump. A clear distinction between the diffusion and ejector principles is needed, but, before considering this, a description of a simple mercury diffusion pump forms a useful introduction to the ways in which a continuous supply of vapour is obtained.

Vapour pumps made of glass are frequently used in the laboratory and can be constructed by a skilled glass-blower, but the modern tendency is to use commercial metal pumps with their considerably better performance.

In all the vapour pumps, the discharge is to a backing space at a maximum pressure of 10^{-1} torr or above, depending on the design. The vapour pump therefore requires to operate relative to a backing pump; the mechanical rotary pump is most frequently used for this purpose. In general, the vapour pump is the choice when it is needed to obtain effective pumping speeds at pressures below 10^{-3} torr, but there is an increasing number of alternatives, especially the getter-ion pumps (section 1.14). So the common vacuum system has a vapour pump backed by a series-connected backing rotary pump.

1.6. *The Mercury Diffusion Pump*

A simple single-stage mercury diffusion pump made of borosilicate glass (Fig. 9) has a discharge outlet to the backing pump; its intake port is connected to the vessel being evacuated. This vessel is first pumped by the backing pump, through this diffusion pump, before the mercury in the boiler is heated. The mercury is then boiled at the backing pressure (10^{-1} torr or below) by seating this boiler in an external electric heater. The mercury vapour creates a pressure of a few torr in the boiler and mercury vapour streams up the central tube T (the pump chimney) to impinge on the reflecting 'umbrella'-shaped cone C. The reflected mercury atoms then travel swiftly, and in quantity, down the annular tube D. The pump nozzle through which this mercury stream emerges is at N. The pump is surrounded by a jacket through which cooling water flows continuously, so that the downward stream of mercury finally condenses against the walls near the bottom of the tube D and returns to the boiler via tube U. This tube is U-shaped so that the liquid mercury collecting in it prevents direct contact between the backing space and the higher pressure in the boiler.

The gas molecules in the space A above the nozzle are moving randomly in all directions with no one direction preferred, in the absence of the mercury vapour stream. However, when the mercury vapour stream is on, below N within the annular space atoms of mercury having velocities of several hundred metre per sec in the downward direction will collide with gas molecules and impart to them velocity components directed towards the backing space region. Gas molecules so directed away from the jet will create voids into which move molecules from the intake port and so from the vessel. This preferred direction imparted to the initially randomly-moving gas molecules establishes a pressure gradient in the pump, whereby the permanent gas pressure in the vicinity of A can readily be much less than that in

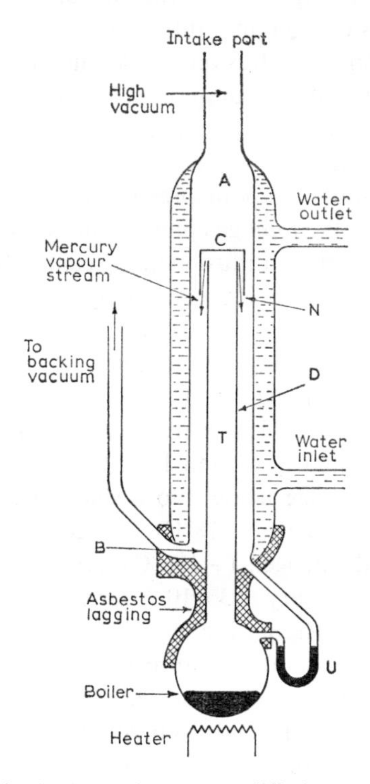

Fig. 9. A simple mercury diffusion pump.

the backing space B. Hence, the vessel connected to the intake port can be evacuated to a permanent gas pressure of 10^{-6} torr or very much lower, where the backing pressure is 10^{-1} torr or below.

The first mercury diffusion pump was made in 1915 by that great genius of vacuum technology, W. Gaede [4]. His pump needed a careful regulation of the mercury vapour temperature (indeed, study of his original pump is convincing evidence of his experimental prowess because one wonders why it worked at all!). But within a year I. Langmuir [5] had introduced the water-jacket to condense the returning mercury and correspondingly stop the tendency of the mercury vapour to stream back into the vessel being exhausted. The temperature of the mercury vapour was thus rendered much less critical; however, the rate of heating the mercury in the boiler is an important factor in determining the backing pressure required for satisfactory operation. Too slow a heating will require a low backing pressure, whereas too fast a heating will send the mercury vapour undesirably into the upper parts of the pump tube to the vessel. To obtain satisfactory condensation, the diffusion nozzle must be inside the water-jacket, and the length of the pump must be great enough to prevent any appreciable amount of permanent gas from diffusing back against the directed mercury vapour stream.

At 20°C, the saturated vapour pressure of mercury is $1{\cdot}2 \times 10^{-3}$ torr. This means that the *total* gas pressure in the vessel connected to the intake port of a mercury diffusion pump cannot be less than about 10^{-3} torr, unless a technique is used to prevent even minute amounts of mercury vapour reaching the vessel. The permanent gas pressure (i.e. due to non-condensable gases) can, nevertheless, be 10^{-6} torr or much lower. The usual remedy is to install a cold trap between the vessel and the intake port of the mercury diffusion pump. Designs of cold trap are considered in section 1.11. Cold traps may be cooled with solid carbon dioxide made into a sludge with acetone or trichlorethylene (temperature -78°C), or cooled with liquid air (-183°C) or liquid nitrogen (-196°C). At -78°C, the saturated vapour pressure of mercury is 3×10^{-9} torr, which appears satisfactory, except that water vapour is almost invariably present during the pumping and has a vapour pressure of 5×10^{-4} torr at -78°C. Therefore, liquid air is much preferred, though liquid nitrogen is superior, because liquid air tends to become primarily liquid oxygen and thus there is a risk of fire and violently explosive reactions with oils and greases. At -183°C, the vapour pressure of mercury is estimated to be some 10^{-21} torr, and that of water is also vanishingly small.

1.7. *Outline of the Theory of Action of the Vapour Pump: More Advanced Designs*

In the vapour pump, the essential action is that a stream of the pump fluid vapour emerges from a nozzle and expands into a pump casing (initially evacuated to a backing pressure), on the cooled walls of which it is condensed. This vapour, streaming in the direction towards the pump discharge outlet, at backing pressure, imparts momentum in this direction to the gas molecules, so creating a pumping action and establishing a pressure gradient in the gas between the intake aperture at high vacuum and the discharge outlet at backing pressure. In addition, this vapour stream creates a 'seal' across the

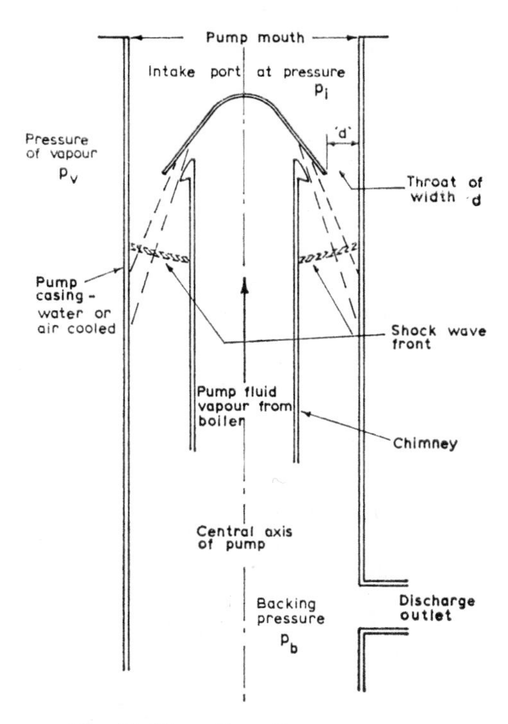

Fig. 10. The action of a vapour pump.

pump which prevents gas molecules from diffusing from the backing region back towards the intake aperture. Fig. 10 illustrates the basic features of a vapour pump.

It is necessary to establish that momentum can be imparted to the gas molecules in the required direction, so as to cause pumping, and also that the 'seal' effect across the pump is obtained.

Both these requirements demand that the velocity of the molecules in the vapour stream emerging from the nozzle exceeds the average molecular velocity determined by the temperature of the vapour before emerging from the nozzle.

The average velocity of a molecule at absolute temperature T is given by equation (1.7) as

$$\bar{v} = \sqrt{\left(\frac{8RT}{\pi M}\right)}$$

But $pV = RT$ and $\rho = M/V$, where p is the pressure, T the absolute temperature, ρ the density, M the molecular weight, and V the volume of 1 gram-molecule of the vapour.

$$\therefore\ \bar{v} = \sqrt{\left(\frac{8p}{\pi\rho}\right)} \tag{1.11}$$

The speed of sound, v_s, in a gas or vapour at pressure p is given by the well-known equation

$$\bar{v}_s = \sqrt{\left(\frac{\gamma p}{\rho}\right)} \tag{1.12}$$

where $\gamma = C_p/C_v$, C_p being the specific heat at constant pressure and C_v the specific heat at constant volume.

From equations (1.11) and (1.12), it follows that

$$\frac{\bar{v}}{v_s} = \left(\frac{8p/\pi\rho}{\gamma p/\rho}\right)^{1/2} = \left(\frac{8}{\pi\gamma}\right)^{1/2}$$

For a monatomic vapour, such as mercury, $\gamma = 1{\cdot}67$; for a diatomic molecule, $\gamma = 1{\cdot}4$.

In the case of mercury therefore

$$\frac{\bar{v}}{v_s} = \left(\frac{8}{1{\cdot}67\pi}\right)^{1/2} = 1{\cdot}24$$

The speed of sound in a gas or vapour is hence of comparable magnitude to the average molecular velocity. The velocity of the vapour stream thus needs to be supersonic. Otherwise, if the vapour supply emerged from the nozzle with the average kinetic velocity decided by its temperature, it would simply spread out inside the pump casing without imparting any considerable directional momentum to the gas molecules. Then the pumping speed would be very small.

To ensure a satisfactory vapour stream, the pump fluid in the pump boiler has to be heated until the pressure of its vapour, p_v, is considerably higher than the backing pressure, p_b, provided in the pump (Fig. 10) and so also higher than the gas pressure, p_i, to be created at the intake port. The stream of vapour emerging from the nozzle consequently expands into a region at lower gas pressure; in doing so, it will expand adiabatically and can be shown to acquire supersonic velocities with a satisfactory rate of discharge provided

$$\frac{p_c}{p_v} = \left(\frac{2}{\gamma+1}\right)^{\gamma/(\gamma-1)} \tag{1.13}$$

where p_c is the **critical back pressure,** p_v is the initial pump fluid vapour pressure before expanding through the nozzle, and γ is the ratio C_p/C_v for the pump fluid.

For example, for mercury, $C_p/C_v = 1{\cdot}67$, so

$$\frac{p_c}{p_v} = \left(\frac{2}{2{\cdot}67}\right)^{1{\cdot}67/0{\cdot}67} = 0{\cdot}75^{5/2} = 0{\cdot}49$$

As the supersonic stream of pump fluid vapour travels through the gas in the pump, the vapour molecules collide with and entrain gas molecules in the mixing region within the pump casing. These gas molecules therefore have imparted to them marked velocity components directed towards the discharge outlet to the backing pump, establishing a pressure gradient and a pumping action.

The supersonic high pressure region of vapour overtakes the slower moving gas molecules which have sonic speeds. There is consequently a pressure rise resulting in a steep and stable wave front, i.e. a shock wave is formed in which the gas is rapidly compressed. This shock wave will occur at a distance from the nozzle which increases with decrease of the backing pressure p_b (Fig. 10). This shock wave acts like a 'dam', providing the 'seal' required across the pump. Gas from the backing region cannot surmount the pressure increase in this shock wave and reach the high vacuum intake aperture. If the backing pressure is too high, the shock wave front will be too near the nozzle outlet. The sealing action is then less effective; gas from the backing region will be able to 'back-diffuse' to the vapour pump intake aperture, especially in the region near the cooled pump walls, so increasing the ultimate pressure.

Depending on its design, a vapour pump therefore has a **critical backing pressure,** p_{cb} (not to be confused with the critical back pressure, p_c), above which there is a more or less sudden increase of the pressure on the high vacuum side of the pump. The critical backing pressure is lower than the critical back pressure (a value of 0·05 to 0·1 torr being usual), but it will depend considerably on the gas being pumped and the pump design.

A well-known result from Knudsen's study of the kinetic theory of gases is that the number of gas molecules which impinge on unit area of a boundary in the gas per sec, N, is given by equation (1.8)

$$N = \tfrac{1}{4}\, n\bar{v}$$

where n is the number of molecules per unit volume and $\bar{v}$ the average molecular velocity. If, therefore, A is the area of the intake aperture of the vapour pump, i.e. the throat area which is the annular area between the pump casing and the nozzle (Fig. 10), the number of gas molecules entering the pump per sec would be $\frac{1}{4}n\bar{v}A$ if the pump were perfect, i.e. there were no impedance to flow of gas through the intake aperture and all the molecules of gas which hit the aperture passed through the pump without any return. In practice such perfection is not possible; the actual rate of removal of gas molecules is $\frac{1}{4}n\bar{v}AH$, where H is a **speed factor** defined as the ratio of the actual speed to the speed if the pump were perfect. Thus

$$H = S_a/AN$$

where S_a is the actual pump speed in gas molecules per sec. This actual speed depends on the nature of the gas and the pump design. For a good design, H will be about 0·35 for nitrogen.

Clearly, high pumping speeds are only obtainable with large intake throat areas. The ideal demand is thus a very wide aperture pump with a very small nozzle. However, the pump fluid vapour from the nozzle must not only have high molecular concentration and supersonic speed, but also be able to spread towards the cooled walls after leaving the nozzle, if the necessary seal effect provided by the shock wave front is to be effective. Otherwise, back-diffusion of gas from the fore-region along the casing walls to the high vacuum region will occur. In practice, satisfactory performance demands a ratio between the area of the expanded jet and the nozzle outlet of 5 to 10. The necessity for a large intake throat, suitable vapour stream, and a suitable critical backing pressure is best met in practice by a **multi-stage vapour pump**, where two, three, or four stages are used; the second stage backing the first top-stage, and the third stage the second, as shown in Fig. 11. By this means, higher critical backing pressures, as provided by the mechanical pump, are allowable; up to 0·5 torr for an oil vapour pump, and 1 torr in mercury vapour pumps, with much higher values of 30 torr or more in diffusion-ejector designs.

The vapour diffusion pump is designed to have a pumping speed which is constant below about 10^{-3} torr down to the ultimate pressure provided, which can be in the ultra-high vacuum region at 10^{-9} torr, or below if the design is suitable. Indeed, theoretically, there is no limit to the ultimate pressure obtainable, but stringent observation of practice is demanded to obtain ultimates below 10^{-8} torr (section 3.8).

The vapour ejector pump, on the other hand, though having a similar working principle, is designed to have a maximum pumping speed at intake pressures of 2×10^{-2} to 10^{-1} torr and above, and to

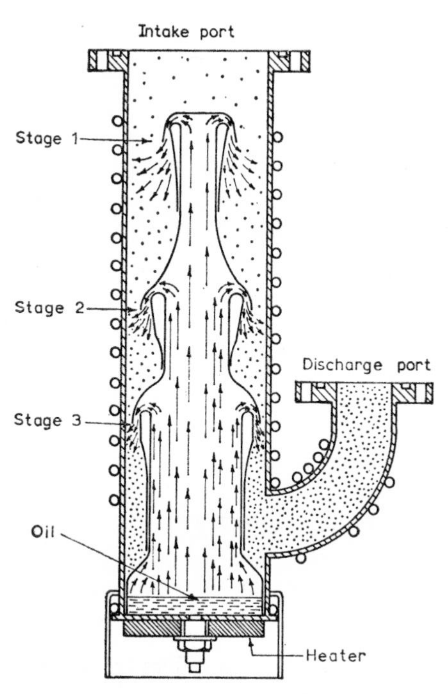

Fig. 11. A three-stage oil diffusion pump. (Dots indicate permanent gas.)

operate at a backing pressure of 0·5 to 1·0 torr with oil ejectors, and, in the case of mercury ejectors, up to 5 or even 100 torr.

In the diffusion pump, the m.f.p. of the gas molecules at the intake aperture is some 50 cm or longer as intake pressures are 10^{-4} torr and below (equation 1.4). Collisions of the gas molecules with the walls are therefore more important than intermolecular collisions because this m.f.p. is larger than the pump throat diameter. Indeed, the gas flow is **molecular** (section 3.1) and the pumping speed is largely independent of the intake pressure below 10^{-3} torr (Fig. 12*a*). In the ejector pump, however, the intake pressure is generally about 1,000 times greater than in the diffusion pump, and the m.f.p. of the gas molecules is only about 5×10^{-2} cm. Intermolecular collisions are now more important than collisions with the walls: the gas flow is chiefly **viscous** (section 3.1) and the pumping speed varies considerably with pressure showing a marked peak value around a particular pressure (Fig. 12*b*).

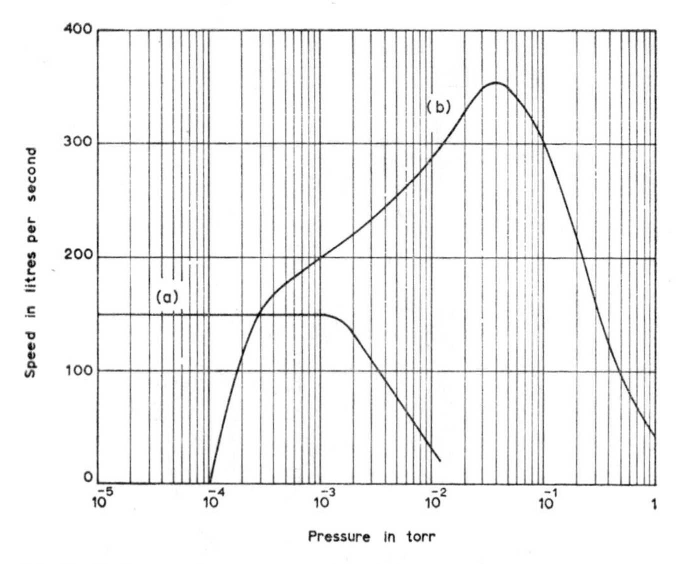

Fig. 12. Typical pumping speed against intake pressure characteristics for: (*a*) an oil vapour diffusion pump; and (*b*) an oil vapour ejector pump.

The design of the ejector pump differs from that of the diffusion type, because a dense vapour stream with supersonic speed has now to be produced from a nozzle entering a mixing region at a higher initial backing pressure of 1 torr or more. Considerably higher boiler pressures are thus demanded to provide a sufficiently large value of p_v (equation 1.13). As this vapour stream is surrounded by gas at comparatively high pressures of the order of 10^{-1} torr in the nozzle region, its spread is much more limited than in the diffusion pump. To ensure a satisfactory 'seal' due to the shock wave, a convergent water-cooled pump casing is used in the mixing region, so that satisfactory condensation of the directed vapour is obtained and back-diffusion of gas to the intake port is prevented. The vapour stream now entrains the gas by viscous drag rather than by diffusion and conveys it at high speed to the discharge outlet to the backing pump (Fig. 13).

The ejector pump is valuable for exhausting large systems at high speed where pressures in the range from 10^{-1} to 10^{-4} are to be established. It has been introduced, indeed, especially as an oil vapour ejector, primarily to provide high pumping speeds in the region below 10^{-1} torr, where the speed of a mechanical pump begins to decrease, and above 10^{-3} torr, where the speed of the diffusion pump falls off.

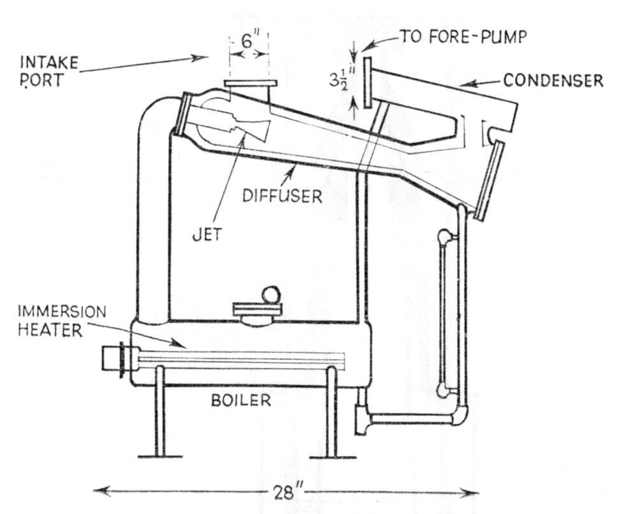

Fig. 13. An oil vapour ejector pump.

In 'filling the gap' in the pressure range between 10^{-1} and 10^{-4} torr, the chief rival to the ejector pump is the Roots pump (section 1.13).

The ejector pump is often used as an intermediate between a diffusion pump and a mechanical backing pump, a practice frequently adopted when large quantities of gas have to be handled at intake pressures of 10^{-4} torr and below. In several designs, the ejector stage is incorporated within the single vapour pump giving a multi-stage diffusion-ejector pump (Fig. 14), in which the first stage is a diffusion nozzle or jet, the second stage is probably partly diffusion and partly ejector, and the third and any further stages are ejector.

For fuller discussions of the theory of the working of vapour pumps see: Alexander [6]; Florescu [7]; Jaeckel [8]; Jaeckel *et al* [9]; and Nöller [10].

1.8. *Pump Fluids: Mercury and Oils*

The operating fluid used in either an oil-sealed mechanical pump or a vapour pump is known as the **pump fluid**. A high purity hydrocarbon, with multi-functional additives to impart anti-rust and anti-lacquering properties with resistance to oxidation and acids, is generally used for mechanical pumps. The vapour pressure at 20°C is about

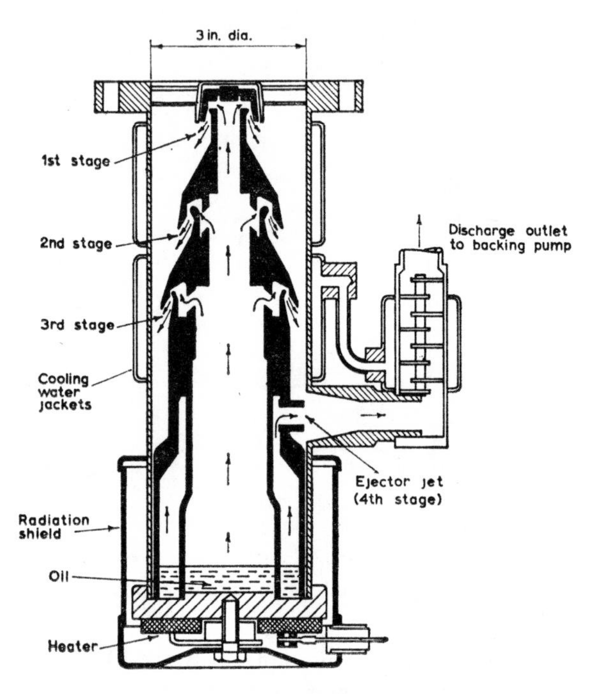

Fig. 14. An oil diffusion-ejector pump (Edwards High Vacuum Ltd., Model EO2).

10^{-3} torr and the oil must have non-corrosive, lubrication and viscosity characteristics suitable for the pump. As there are a wide variety of pumps from the various manufacturers, the best practice is to use the oil recommended in the pump manufacturers' catalogues.

The pump fluid for a vapour pump has to be chosen in relation to the performance expected and the operation of the pump, but bearing in mind that the best fluids are expensive. The vapour diffusion pump is usually required to have a high pumping speed per unit area of its intake aperture and be capable of producing a low **ultimate pressure** (which is the limiting pressure attainable after a pumping time long enough to ensure that further reductions in pressure in the chamber are negligible). Further, it should work against a critical backing pressure which should not need to be too low. To provide good entrainment of gas molecules by the pump fluid vapour stream and ensure the necessary seal-off effect due to the shock wave, a pump fluid of high molecular weight is an advantage. Mercury has an atomic weight of 201; all the oils used have molecular weights between 300 and 500. As regards the ultimate pressure, this depends on: whether or not a baffle and/or cold trap (section 1.11) are used; the seals used in the system; whether the chamber is baked or not, and from what material the chamber is fabricated; leaks in the system; and the design of the vapour pump in relation to purification of its own working fluid. In general, the pump fluid should have as low as possible a vapour pressure at the temperature of the intake aperture. At the same time, it should have a vapour pressure against temperature curve such that boiler pressures of 0·5 torr or more can be achieved without excessive boiler temperatures. If an excessive boiler temperature is used, the pump fluid may break down into fractions, some of which can readily be non-condensable gases which back-diffuse to the high vacuum side of the pump. Thus, a vapour pressure of 10^{-9} torr at 20°C is demanded for an ultra-high vacuum system where a vapour diffusion pump is to be used without a cold trap. This is an excessive and expensive demand, not required if an ultimate pressure of 10^{-5} torr is adequate or cannot be improved upon because of actual leakage or virtual leakage due to vapour from an unbaked chamber, seals, valves, etc. Again, the prime requirement for the fluid in a vapour ejector pump is not low vapour pressure but resistance to disintegration at high boiler temperatures.

Further requirements of vapour pump fluids are: (*a*) a suitable viscosity against temperature characteristic, to ensure ready flow back to the boiler at the cooling water temperature; (*b*) chemical stability against temperature rise and in the presence of the common metals, glass and gasket materials; (*c*) resistance to oxidation when heated at atmospheric pressure; and (*d*) resistance to decomposition on exposure of the vapour to hot filaments or to a gaseous discharge.

Mercury has two considerable virtues as a pump fluid: its density and viscosity ensure ready circulation in the pump; and it is an element which cannot decompose on heating or in the presence of hot filaments or an electrical discharge. These latter characteristics lead to the choice of mercury vapour pumps for three main applications:

(*i*) for ejector pumps operating at high boiler pressures and able to work against a backing pressure of 30 torr or more:
(*ii*) in ultra-high vacuum systems for gas analysis where the presence of mercury vapour is readily distinguished by a mass spectrometer from other gases;
(*iii*) for pumping particle accelerators and high voltage equipment in which oil contamination may affect targets, or cause insulation breakdown.

The three chief difficulties with mercury are:

(*i*) it has a vapour pressure of about 10^{-3} torr at 15°C (Table 1.2) and so cannot produce a lower total pressure than this in a chamber, unless a cold trap (section 1.11) is used between the intake aperture of the pump and the chamber;
(*ii*) it reacts readily with several materials and, in particular, amalgamates with copper, brass, and aluminium, so these metals must not be allowed to 'see' the mercury vapour in a system;
(*iii*) the pumping speed and the back-streaming (section 1.11) of a mercury vapour pump are greatly affected by small amounts of hydrocarbon contamination.
It should also be remembered that mercury vapour is toxic.

The pioneers in the introduction of low vapour pressure oils suitable for vapour pumps were Burch [11] and Hickman [12]. Of the oils listed in Table 1.3, the silicone oils (which are semi-organic polymers) are the most widely used in high vacuum technique.

It is not good practice to use a vapour pump designed to be employed with mercury and substitute an oil. Indeed, it is ill-advised to use other than the specific oil recommended by the manufacturer for a particular vapour pump. Reasons for this are:

(*i*) with an ejector or a diffusion-ejector pump, the temperature may cause severe decomposition of the oil;
(*ii*) the nozzle dimensions and the boiler temperature are critically adjusted to the fluid used to obtain optimum performance in relation to speed, ultimate pressure and back-streaming;
(*iii*) the temperature distribution in the pump needs to be related to the viscosity-temperature characteristic of the pump fluid to ensure satisfactory circulation;
(*iv*) the backing pressure tolerance depends on the shock wave seal-off effect provided by the vapour stream, which is decided by the

TABLE 1.2

Vapour pressure in torr of mercury at various temperatures in °C

°C	−180	−78	−30	−20	−10	0	10
torr	negligible	3×10^{-9}	$4{\cdot}8 \times 10^{-6}$	$1{\cdot}8 \times 10^{-5}$	6×10^{-4}	$1{\cdot}85 \times 10^{-4}$	$4{\cdot}9 \times 10^{-4}$
°C	20	40	60	80	100	150	200
torr	$1{\cdot}2 \times 10^{-3}$	6×10^{-3}	$2{\cdot}5 \times 10^{-2}$	$8{\cdot}9 \times 10^{-2}$	$2{\cdot}7 \times 10^{-1}$	2·8	17·3

speed of this stream, which in turn is determined by the temperature and the molecular weight of the oil.

The suppliers of the oils listed in Table 1.3 are as follows (the numbers given are those in the extreme left-hand column of the table): 1, 2, and 3 – the Shell Chemical Co., Ltd. (U.K.) and James G. Biddle Co. (U.S.A.); 4, 5, 9, and 10 – Consolidated Vacuum Corp. (U.S.A.); 6 – Litton Engineering Laboratories (U.S.A.); 7 and 8 – National Research Corp. (U.S.A.) and Vacuum Industrial Applications (U.K.), who use the name 'Viacoil' instead of Narcoil; 11, 12, 13, and 14 – Dow-Corning Corp. (U.S.A.) and Midland Silicones Ltd. (U.K.), for whom the agents are Edwards High Vacuum Ltd.

Oils 1, 2, 3, 5, and 6 are moderately-priced general-purpose fluids, but do not resist oxidation. Apiezon C and also the Octoils (9 and 10) are often used in glass-fractionating pumps and metal self-purifying pumps. The silicone oils (11 to 14) are semi-organic polymers which are exceptionally resistant to oxidation at high temperatures compared with all the other oils except the Narcoils (7 and 8) and Convalex (4). As a result, heating of these oils at atmospheric pressure does not seriously impair their efficiency as pump fluids, though it is certainly undesirable practice. They do not oxidize even on exposure to air at pump boiler operating temperatures. Such drastic treatment of the other oils listed in Table 1.3 results in the formation of tars and corrosive vapours in the pump, which block up nozzles and stick to the surfaces of the nozzle system so reducing noticeably the pumping efficiency. For this reason, the silicones are the most popular pump fluids for vapour diffusion pumps. Latham, Power, and Dennis [13] performed a series of trials on vapour pump fluids in which an ultimate pressure of about 5×10^{-6} torr was established by a vapour diffusion pump filled with one or other of the fluids, and then air was admitted at atmospheric pressure whilst the pump heater was left on. Regular repetition of this brutal treatment several hundred times, over a period of many days, established clearly that the pump interior assembly was still clean if the silicone oils or chlorinated diphenyls (Narcoils) were used, but other fluids, such as the Apiezon oils and Octoils, gave rise to tarry and solid carbonaceous deposits on the nozzle system which severely depreciated the pump performance.

Pump fluids give difficulty due to back-streaming, whereby some of the oil will enter the chamber being pumped (section 1.11). Several cases have been recorded where in experimental work on electron tubes, mass spectrometers, ion accelerators, etc, small quantities of

TABLE 1.3

Oil vapour pump fluids

	Fluid		*Vapour pressure in torr at 25°C*	*Temperature in °C for a vapour pressure of 10^{-2} torr*	*Specific gravity at 25°C*
1	Apiezon A	Mixture of hydrocarbons	2×10^{-5}	110	0·8735 at 15°C
2	Apiezon B	Mixture of hydrocarbons	4×10^{-7}	127	0·871 at 15°C
3	Apiezon C	Mixture of hydrocarbons	10^{-8}	160	0·880 at 15°C
4	Convalex 10	Polyphenyl ether	10^{-10}		
5	Convoil 20	Hydrocarbon	5×10^{-7}		0·865 at 15°C
6	Litton oil	Hydrocarbon	$1{\cdot}5\times10^{-7}$	132	
7	Narcoil 10	Chlorinated diphenyl	3×10^{-4}		1·54
8	Narcoil 40	Di-trimethyl hexyl phthalate	10^{-7}		0·973
9	Octoil	Di-2-ethyl hexyl phthalate	3×10^{-7}	123	0·9796
10	Octoil S	Di-2-ethyl hexyl sebacate	3×10^{-8}	144	0·9103
11	Silicone DC 702	Mixture of polysiloxanes	10^{-6}	115	1·07
12	Silicone DC 703	Mixture of polysiloxanes	10^{-7}	145	1·09
13	Silicone DC 704	Single molecule materials	4×10^{-8}	215 at 0·5 torr	1·07
14	Silicone DC 705	of semi-organic nature	4×10^{-10}	200 at 0·3 torr	1·09

pump oil back-streaming into the chamber have become deposited on electrodes, and then on heating or electron or ion bombardment the surfaces have become coated with tenacious films. These films tend to be non-conducting silicaceous compounds if silicone pump fluids are used, and more troublesome than the conducting carbonaceous deposits resulting from the use of hydrocarbon pump fluids such as the Apiezon oils.

The range of silicone oils (in which DC 702 and 703 are general purpose vapour diffusion pump fluids for use down to ultimate pressures 10^{-6} and 10^{-7} respectively) has been extended recently by the introduction of DC 704 (Huntress, Smith, Power, and Dennis [14]) and DC 705. Both of these, unlike DC 702 and 703, are allegedly single component materials having a definite molecule of a single kind. They markedly resist decomposition on heating in the pump boiler under vacuum, and they are also resistant to gamma radiation. Such decomposition of the earlier (but less expensive) oils gives rise to vapours and gases which travel undesirably against the pumping direction into the chamber at the pump intake. Furthermore, there can well be present gaseous components of such low boiling points that they are not condensed in a cold trap even at the temperature of liquid nitrogen. This decomposition is not entirely absent with DC 704 and 705, but markedly less significant than with 702 and 703.

As compared with the element mercury, all oils will decompose to a more or less extent in the presence of hot filaments and in an electrical discharge, aggravating seriously the problem of pressure recording by hot-cathode and also cold-cathode ionization gauges (sections 2.7 and 2.6).

Two of the oils listed in Table 1.3, DC 705 and Convalex 10, have vapour pressures of about 10^{-10} at room temperature. The use of these oils therefore leads to the possibility of producing ultra-high vacua in oil-diffusion-pumped chambers without the necessity for a cold trap between the pump and the chamber. In addition to the avoidance of pumping speed restriction, this possibility leads to enormous saving in running costs with very large plants such as proton-synchrotrons, where the annual cost of liquid nitrogen may amount to £100,000 or more. However, to establish pressures of 10^{-9} and below, the problem of bake-out of very large chambers remains and also that of back-streaming deposits from oils. In this latter connection, Hickman [15] reports that the polyphenyl ethers (such as Convalex 10) can afford positive advantages on back-streaming into the chamber as they may be allowed to coat gaskets, O-rings and the inside walls of the chamber with a deposit which greatly reduces outgassing and the escape of volatiles. Certainly, the problem of setting-up demountable ultra-high vacuum systems of moderate size (e.g. for vacuum coating purposes), where frequent admission of air and repumping are undertaken, and where bake-out is not easily practised, has not yet been satisfactorily solved; and the intelligent use of vapour diffusion pumps with Convalex 10 or DC 705 affords an attractive, promising approach.

In Table 1.3, Narcoil 10 (7) is a useful oil for vapour ejector pumps. Plasticizers such as tri-xylenyl phosphate, Arochlor (Monsanto Chemical Co., U.S.A.) and Edwards High Vacuum Ltd. booster pump fluid A are also useful for this purpose.

1.9. *Pumps Designed to Purify their own Working Fluid*

There are three types of vapour diffusion pump which utilize a working pump fluid in the form of an oil, which, like most of those listed in Table 1.3, is a mixture of components of various molecular weights and volatilities. The object of these designs is to ensure that the first-stage nozzle as far as possible operates primarily with the least volatile component of lowest vapour pressure, and that the more volatile constituents are rejected from the main vapour stream. Such pumps are not as important as they were because of the introduction of single component oils, such as DC 704 and 705.

The first pumps of this kind were glass **self-fractionating oil diffusion pumps** introduced by Hickman [16]. An example is shown in Fig. 15.

This pump utilizes a divergent cone-shaped nozzle, A, (first introduced much earlier by Crawford [17]) as the first stage, and a cylindrical nozzle, B, as the second stage, at higher pressure intermediate between that of the high vacuum to be maintained at the intake port I and the backing vacuum at the discharge outlet D. The three inter-connected pump boilers contain electric immersion heaters in the oil charges. The most volatile constituents of the oil (Apiezon C or Octoil S are often used) are distilled into the alembics in the vertical glass tube connecting the discharge port to the backing pump. The oil constituents of medium volatility form the operative vapour streams at the nozzles A and B, whilst the comparatively non-volatile components and impurities go to the smaller boiler E, where the dark-coloured non-volatile constituent collects and the more active fractions are distilled off and returned to the other two boilers.

The heat inputs to the separate boilers have to be fairly precisely regulated. The construction, whereby the main axis of the pump is sloped slightly from the horizontal, facilitates effective separation of the constituents and enables the nozzle diameters to be determined independently of the boiler areas.

A pump of this type, operating with an oil of nominal vapour pressure at 25°C of 10^{-7} torr, can readily achieve an ultimate pressure (without baffle or cold trap) of 5×10^{-8} torr.

A schematic cross-section in the vertical plane through a large **metal oil diffusion pump of the fractionating type** (Fig. 16) shows the

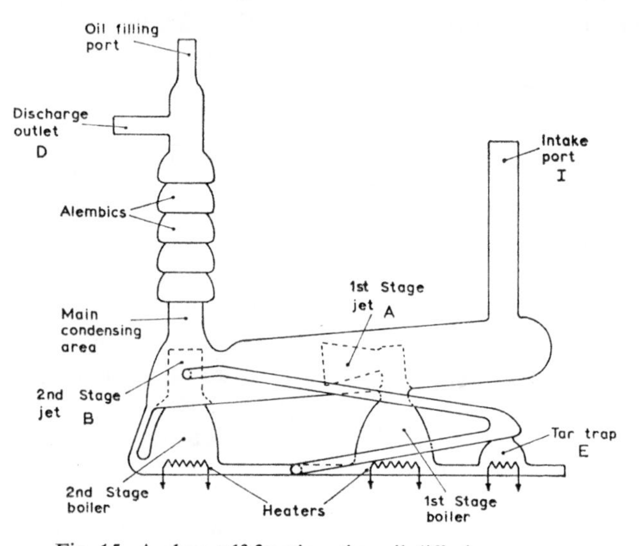

Fig. 15. A glass self-fractionating oil diffusion pump.

essential difference in construction between this pattern and the ordinary non-fractionating pump: that is that each of the stages are supplied with oil vapour from their own cylindrical tube or chimney, whereas a single chimney supplies all nozzles of the normal pump. In the three-stage fractionating pump shown, the oil vapour issuing from the first-stage nozzle 1 is from cylindrical chimney 1, the second-stage nozzle 2 from the annular region 2 between two cylinders, and nozzle 3 from annular region 3. The purifying action on the oil is simply that the condensate from all three nozzles enters first the outer annular section of the boiler, from which the most volatile fractions of highest vapour pressure boil off, to feed the third stage 3, acting as a backing (ejector) stage to stage 2. The top first-stage nozzle, 1, fed from the central compartment of the boiler, thus receives primarily the least volatile constituents of the oil, i.e. those of the lowest vapour pressure. In general, a fractionating oil diffusion-ejector pump will provide an ultimate pressure in an unbaked chamber, without the use of baffles and/or cold traps, of 10^{-7} to 5×10^{-7} torr (though this depends on the gasket materials used and the type of chamber, so this figure is only an approximate guide). Whereas a pump of similar dimensions and speed, but non-fractionating, will have an ultimate some 5 to 10 times greater in the same circumstances.

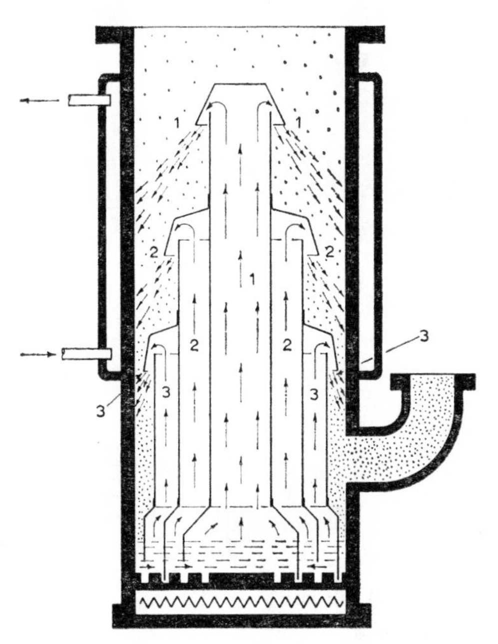

Fig. 16. A metal fractionating oil diffusion pump.

The **self-purifying oil diffusion pump** is represented by the F 203 model of Edwards High Vacuum Ltd. (Fig. 17). In this pump, the oil (Apiezon C is used) vapour emerging from the first- and second-stage jets passes a skirt and falls as a thin film down an elongated pump wall to the boiler. This section of the pump wall is surrounded by a radiation shield, so that it acts as an outer boiler, which supplies oil vapour only to the simple side-stage jet directed straight towards the discharge outlet tube to the backing pump. In falling in the form of a film down the wall, the lighter, more volatile fractions of the oil evaporate first, to feed the side-stage, and the oil which returns finally to the main boiler is of lower vapour pressure, as it is partly freed of these more volatile constituents. It is this main boiler which feeds the first-stage nozzle, where the lowest pressure is required, and also the intermediate second-stage.

1.10. *Air-cooled Vapour Pumps*

Several manufacturers now supply oil diffusion pumps in the smaller sizes, with air cooling by means of fins around the pump body

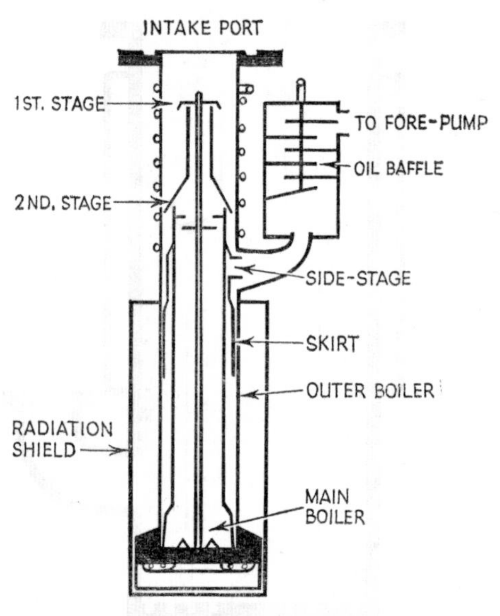

Fig. 17. A self-purifying oil diffusion pump.

through which air is forced by an electrically-driven fan which is an integral part of the assembly. In recent years, the design of the air-cooling system for such pumps has received extra attention, tending to off-set the impression gained from experience that air-cooled pumps did not (despite makers' claims) provide as low an ultimate pressure as their water-cooled counterparts. The convenience of these pumps with their independence of a water supply is obvious in transportable equipment. They are becoming increasingly used in continuously-pumped X-ray crystallography units, some electron microscopes, and other vacuum systems where only an a.c. mains supply is needed to make them fully operative.

1.11. *Back-streaming: Baffles and Traps*

When a vapour pump has a chamber connected at its intake port in a vacuum system consisting basically of chamber, vapour pump, and mechanical backing pump, the probability exists that some of the pump fluid will not travel in the direction (usually downwards) from the nozzles to the backing region of the vapour pump, but in the opposite direction (usually upwards), into the chamber. Indeed, when an oil vapour pump is connected directly to the chamber without any intermediate baffle or trap, its pumping speed is unrestricted, but the collection of pump fluid in the chamber can be so significant as to be readily perceived by eye after some hours' running.

To describe this phenomenon two terms are used: **back-streaming** and **back-migration**. The former is due to molecules of the pump fluid which leave the pump nozzle (particularly that of the top first-stage) in the wrong direction, i.e. with a velocity component opposite to that of the main vapour stream which causes pumping, or which may travel initially in the correct direction but are reflected backwards on impact with a solid surface in the pump near the first-stage jet. The second phenomenon, back-migration, is due to vaporization into the chamber of pump fluid molecules which cling to surfaces within the pump, particularly those of the first stage, the pump walls, and also the surfaces of baffles, traps, or isolation valves used. It is not easy to distinguish between the effects of back-streaming and back-migration: in general, the former is the more important in oil vapour pumps and the latter in mercury vapour pumps. Back-streaming may be reduced by careful nozzle design, but usually reduction by this procedure is at the sacrifice of pump speed. Back-migration is reduced by attention to the nature of the surfaces in the pump near the top-stage jet, and by designing the pump so that these surfaces do not run wet with the pump fluid.

The term **back-diffusion** is sometimes confused with back-streaming. Back-diffusion is a different phenomenon entirely: it signifies that permanent gas or vapour (not the pump fluid) which travels (usually near the inner walls of the pump) from the backing to the high vacuum region.

For an oil vapour pump, back-streaming (including the less important back-migration) is quoted in mg of pump fluid per sq cm of the pump mouth (taken to have a cross-sectional area of πd^2, where d is the diameter of the circular aperture) section reaching the chamber per min. (Its measurement is described in section 4.3.) Alternatively, it is quoted as a total for a given pump of a given mouth diameter. It will depend on the type of pump, the water (or air) cooling, and the temperature of the pump boiler; but is not significantly different for different pump oils if quoted in molecules per min rather than mg per min. It tends to increase significantly at pump intake port (mouth) pressures exceeding 10^{-4} torr, but below this pressure does not vary noticeably with pressure. Exposure of the diffusion pump to the chamber at initial pressures exceeding 10^{-1} torr is reprehensible practice, as temporarily it will greatly increase back-streaming.

For a normal non-fractionating or non-self purifying oil vapour diffusion pump without baffle or trap, the back-streaming rate, with correct heater wattage, and cooling water flow rate as recommended by the manufacturer, will be very roughly 10^{-1} mg per sq cm per min. As most pump oils have a density of near 1 gram per cu cm, the rate quoted in ml instead of mg will be 0·001 times this figure approximately, i.e. about 10^{-4} ml per sq cm per min. It is stressed, however, that these values are quoted here as a guide to the order of quantity; considerable variations in the amounts are experienced.

This does not seem to be a significant figure. However, consider an oil vapour diffusion pump of mouth diameter 4 inch (10 cm) and hence mouth area of 25π sq cm. This will introduce, by back-streaming, $10^{-4} \times 25\pi$ (i.e. 8×10^{-3}) ml of pump fluid into the chamber per min, if no baffle or trap is used. Within 500 min, there accumulates 4 ml of pump fluid, which is easily perceived as a film. Moreover, 100 ml of the pump oil could be lost from the pump in approximately 12,500 min or 200 hour of continuous running if no means were provided to reflect the back-streaming molecules into the pump; and 100 ml is the entire oil charge for a 4 inch pump! In practice, the situation is modified by connecting tubing between the pump mouth and the chamber, because at a distance of $1\frac{1}{2}$ times the mouth diameter above the pump mouth the back-streaming rate is only about 3% of that at the mouth itself.

Thus, for an oil vapour pump, the use of a baffle between the pump mouth and the chamber is in most cases a necessity to reduce back-streaming and return oil to the pump. Apart from loss of oil and the fact that oil accumulating in the chamber renders it unclean, the deleterious effects of back-streaming are the formation of silicaceous or carbonaceous deposits on surfaces within the chamber (section 1.8) and the fact that oil vapour molecules will dissociate in the presence of heated filaments or an electrical discharge.

The oil vapour-ejector pump has a similar back-streaming rate to the oil vapour diffusion pump, but for oil diffusion pumps of the fractionating or self-purifying kinds (section 1.9) the back-streaming rate is about 1/10 of that for the non-fractionating variety.

Mercury vapour pumps also give back-streaming and particularly back-migration, but, as the vapour pressure of mercury at room temperature is 10^{-3} torr approximately, this inevitably forms the chief contribution to the total pressure attained. So a cold trap is absolutely necessary to attain lower total pressures. However, if a cold trap is used above either an oil or a mercury vapour pump, it is most important to realize that back-streaming vapour will condense on the trap's cold surfaces and so represent a loss of charge of fluid in the pump, which is serious after several days' running. To avoid this, it is therefore important, especially with metal pumps of mouth diameter of 2 inch or more, to install a baffle (generally water-cooled) between the pump intake and the cold trap to ensure that the pump fluid is largely reflected back into the pump instead of reaching the cold trap.

Many kinds of baffle have been described in the literature. Fig. 18 shows five widely-used types. Fig. 18(*a*) is a frequently used baffle and isolation valve; with the valve fully open, back-streaming is reduced by about 100 times; partially closing the valve reduces back-streaming further but at the sacrifice of pump speed. Fig. 18(*b*) is a simple copper disc supported within the pump mouth and cooled only by conduction through the metal rod support to the water-cooled pump wall. Power and Crawley [18] report that such a simple baffle will reduce the back-streaming rate by more than 3,000 times, i.e. reduce it to about 3×10^{-5} mg per sq cm per min or 3×10^{-8} ml per sq cm per min. This simple device will therefore ensure inappreciable oil charge loss over several months of running. In this connection, the use of a length of tubing between the pump mouth and the chamber will help significantly, but nothing like so well as the disc of Fig. 18(*b*). The use of this disc will, however, inevitably reduce the speed of the pump to about 30 to 40% of its unbaffled value. Figs. 18(*c*), (*d*), and (*e*) are all water-cooled baffles in which the cooling water inlet is to the baffle *before* being passed through the normal spiral or jacket for cooling the pump walls. Fig. 18(*c*) is a useful cup-shaped baffle for an oil diffusion pump of up to 3 inch diameter mouth. Figs. 18(*d*) and (*e*) show more sophisticated and effective baffles valuable for the larger diffusion pumps, of the chevron and Z types respectively. The design of these is based on the premise that there should be no optical path between the pump nozzle and the chamber, so that back-streaming molecules travelling in straight lines at low intake pressures must inevitably make one collision or more with a water-cooled surface before being able to reach the chamber. On collision, the important factor is the **condensation coefficient**, i.e. the number of molecules which are condensed on the surface divided by the number incident in a given time. An optically-opaque baffle would be perfect if this condensation coefficient were unity. The chevron type, for which the

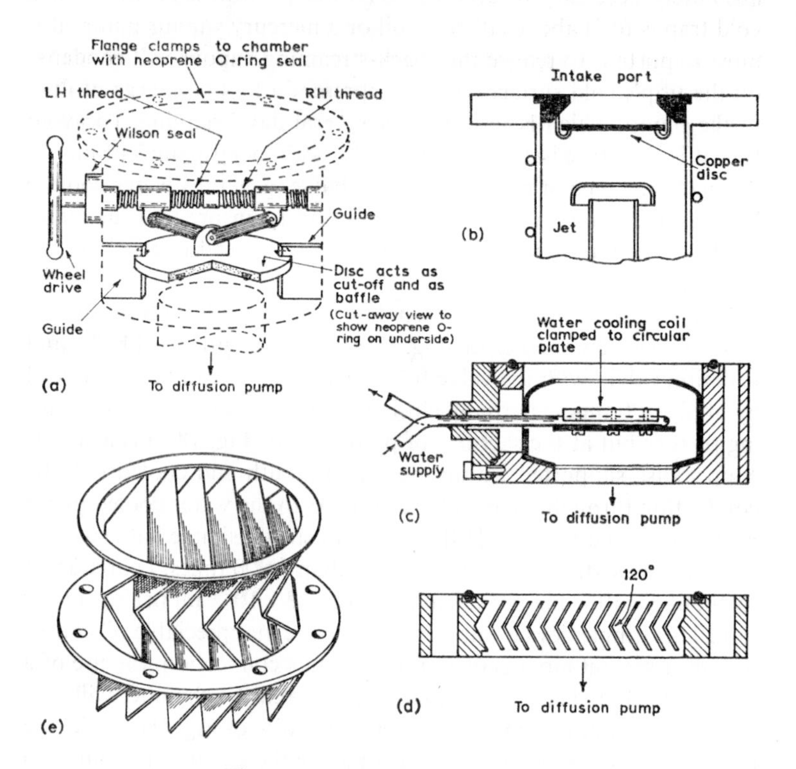

Fig. 18. Baffles for vapour pumps.

optimum angle of the V-shaped slats shown is 120°, is very conveniently shallow in construction so that the pump stack consisting of vapour pump plus chevron baffle is kept short in height; but it is not so efficient as the Z-baffle in which each back-streaming molecule must perforce make two hits, increasing the probability of condensation. The condensation coefficient will approach closer to unity as the baffle temperature is decreased and, moreover, the vapour pressure of the condensed fluid will be greatly lowered (a decrease from 20 to 0°C causes a vapour pressure reduction of about 10 times with most oils). Therefore, baffles of the chevron and Z types are often refrigerated by the circulation of freon from a mechanical refrigerator able to achieve temperatures of −20 to −40°C. A refrigerated baffle is then also a trap. It will reduce back-streaming and back-migration and will also pump by sorption those condensable vapours for which the critical temperature exceeds the baffle temperature.

A baffle should be designed for ready removal to facilitate periodic cleaning.

A convenient but rather expensive means of obtaining a refrigerated baffle is by making use of thermoelectric cooling. The familiar

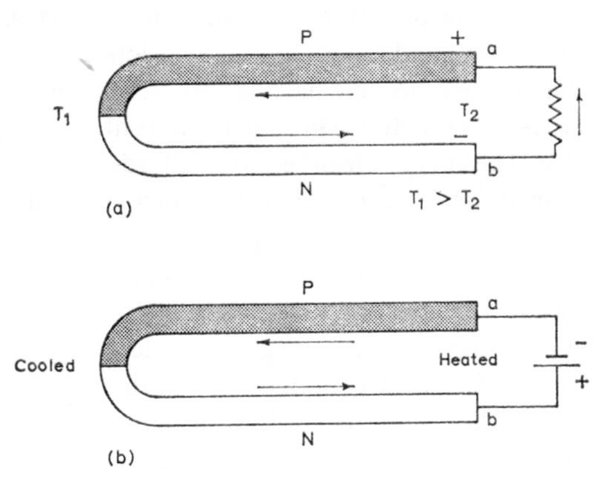

Fig. 19. (*a*) Seebeck effect in a junction between semiconductors. (*b*) Peltier effect in a junction between semiconductors.

Seebeck and Peltier thermoelectric effects are very much larger in magnitude in semiconductors than in metals. For thermoelectrically-cooled baffles, bismuth telluride, Bi_2T_3, is used because of its high thermoelectric power and low thermal conductivity. Consider a rod of an n-type semiconductor (e.g. bismuth telluride containing suitable doping material), N, and a rod of a p-type semiconductor (again, doped Bi_2T_3), P, to be joined at their left-hand ends, while their right-hand ends are connected across a resistive load (Fig. 19*a*). To examine the Seebeck effect, let the left-hand junction be heated to a temperature T_1, and let the right-hand ends across the load be at temperature T_2; where $T_1 > T_2$. Electrons will leave the hot end of a conductor or semiconductor more readily than they leave the cold end. The hot junction of N (in which electrons are the majority carriers) will therefore lose electrons to the cold end more rapidly than electrons leave this cold end. Depending on the temperature difference, $T_1 - T_2$, an equilibrium will be reached between the flow from hot to cold and the return flow, and the cold end will become negatively charged. In the bar P (in which positive holes are the majority carriers), the flow will be in the opposite direction, and point a will be left at a positive potential with respect to b, current flowing through the load (the electron current flow being in the direction indicated on the diagram). In the Peltier effect (Fig. 19*b*), the resistive load is absent and current is passed through the thermoelectric cell from an external source of direct current instead of external heat being applied. With the source connected so that point a is negative with respect to b and therefore the electron current flow direction the same as in Fig. 19(*a*), heat will be absorbed at the left-hand junction, thus it becomes cooled, whereas the right-hand ends become heated. By reversing the current direction, the left-hand junction can be made to heat up and the right-hand end cooled. The energy absorbed (or evolved) in joule at a junction when 1 amp flows for 1 sec defines the Peltier coefficient, π, in joule per coulomb.

The 4 inch thermoelectric baffle of Edwards High Vacuum Ltd. is representative of their range of 2 to 9 inch models. The baffle itself is of the conventional chevron type in a stainless steel body. Cooling of the baffle plates is effected by conduction from them along a copper rod to a set of bismuth telluride cooling elements mounted outside the body of the valve, i.e. outside the vacuum. A current from a d.c. power pack providing 2·5 volt at a maximum of 30 amp is passed through the thermoelectric cooling elements, which are mounted to be electrically in series but thermally in parallel. The maximum temperature difference obtainable is about 45°C. If the hot face of the elements is maintained at cooling-water temperature of 15°C, the chevrons can be cooled to −25°C. Reversing the current direction conveniently enables the baffle to be heated for defrosting. Crawley and Miller [19] describe a compact assembly of such a thermoelectric baffle on a 4 inch, air-cooled, diffusion pump filled with silicone 705 oil. The heat sink is at about 35°C, provided by an arrangement of fins which lie in the air stream which has passed through the upper pump fins. The chevron temperature attained is therefore about −5°C. With a pump of speed 600 litre per sec for air, which is reduced to 200 litre per sec because of the impedance of the baffle, a pressure of 10^{-8} torr was attained in $1\frac{1}{2}$ hour and an ultimate of 10^{-9} torr after an overnight run, without bake-out.

A thermoelectric chevron baffle with a water-cooled heat sink is shown in Fig. 20(*a*) and the principle of thermoelectric cooling is illustrated by Fig. 20(*b*).

Another type of baffle, useful for oil vapour pumps but of no use for mercury vapour ones, makes use of the provision, between the pump and the chamber, of a sorbent material with very large effective surface area for oil vapour. Following the design of Biondi [20], a baffle making use of activated alumina or molecular sieve pellets placed within trays in a simple baffle plate array is illustrated by Fig. 21. This molecular sieve material has an effective surface area of about

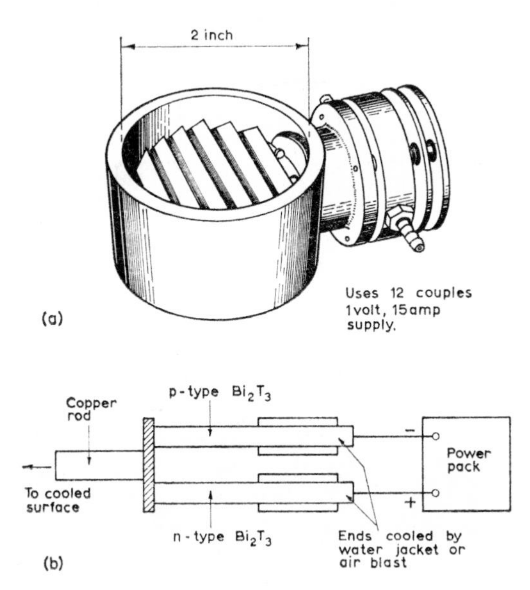

Fig. 20. (*a*) A thermoelectrically-cooled chevron baffle. (*b*) The principle of thermoelectric cooling.

7×10^6 sq cm per gram, so that oil molecules reaching the pellets have a high probability of being retained by physical surface forces. Before use, the molecular sieve pellets are freed of water vapour by heating to 150 to 300°C, (but not above 500°C, as this may destroy their crystallite formation) and are stored in tightly-closed containers. Normally, they are used in baffles at room temperature, but may also be refrigerated to increase their efficiency.

Cold traps may take the form of the refrigerated baffles already described, but the term usually denotes a device introduced between the chamber and the vapour pump intake port in the form of a trap of which the walls are cooled with solid carbon dioxide (−78°C) or liquid air (183°C) or, best of all, liquid nitrogen (−196°C). Indeed, the cold trap also acts as a pump for condensable vapours, and so overlaps in its function with that of the cryogenic pump or cryopump. However, the latter term is reserved here for pumps utilizing surfaces cooled by liquid hydrogen or liquid helium able to sorb gases considered 'permanent' at room temperature (section 1.16).

A cold trap is essential if a mercury vapour pump is used, and if it is required to reduce the total pressure in the chamber to below the

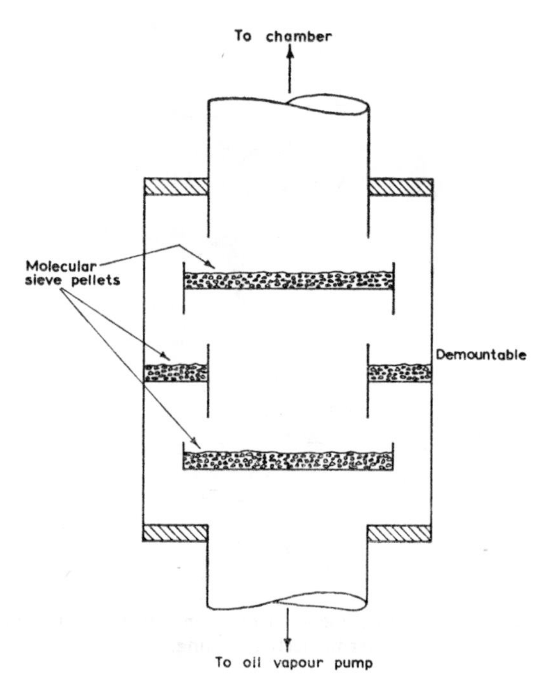

Fig. 21. A molecular sieve baffle for use above an oil vapour diffusion pump.

vapour pressure of mercury at room temperature. Liquid-nitrogen-cooled cold traps are also becoming widely used with oil vapour diffusion pumps in ultra-high vacuum systems (section 3.8).

Fig. 22 shows some designs of glass cold trap. Types (*a*) and (*b*) are immersed in a Dewar flask filled with the liquid coolant. The first is suitable for inclusion in the connection to a McLeod gauge (section 2.3), to reduce to an insignificant level the vapour pressure within the system due to the gauge mercury; and the second is widely used in general-purpose, glass, laboratory systems with a mercury diffusion pump for attaining pressures down to 10^{-7} torr (section 3.6). Fig. 22(*c*) is a re-entrant design in which the central spherical vessel contains the liquid coolant. It forms a convenient alternative to (*b*) but provides a more difficult fabrication problem for the glass-blower.

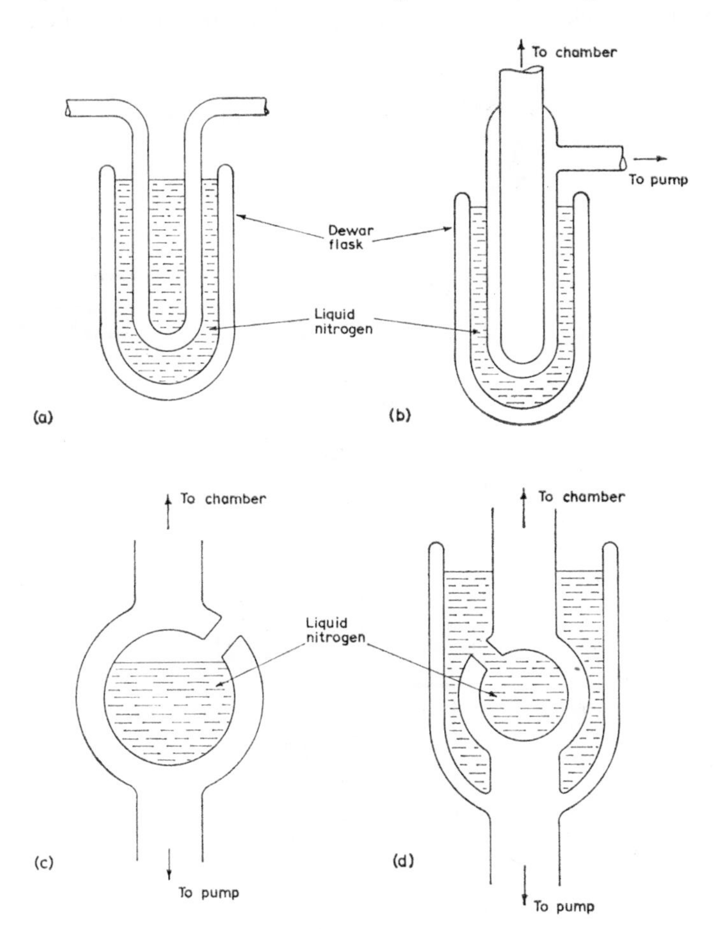

Fig. 22. Glass cold traps.

Fig. 22(*d*) shows an excellent design due to Venema and Bandringa [21] in which, as compared with (*c*), both glass walls, between which the mercury or oil vapour passes, are effectively cooled to −196°C by filling with liquid nitrogen.

Metal cold traps, best made of Inconel or 18/8 stainless steel, are now widely used not only on large metal oil and mercury vapour diffusion pumps but also on smaller pumps (mouth diameter of 2 to 3 inch) where ultra-high vacua are required in a bakeable system (section 3.8). Fig. 23 shows: (*a*) a convenient general purpose design; (*b*) a model offering the attractive facility that it can be readily dismantled for cleaning; and (*c*) a design with an anti-creep skirt due to

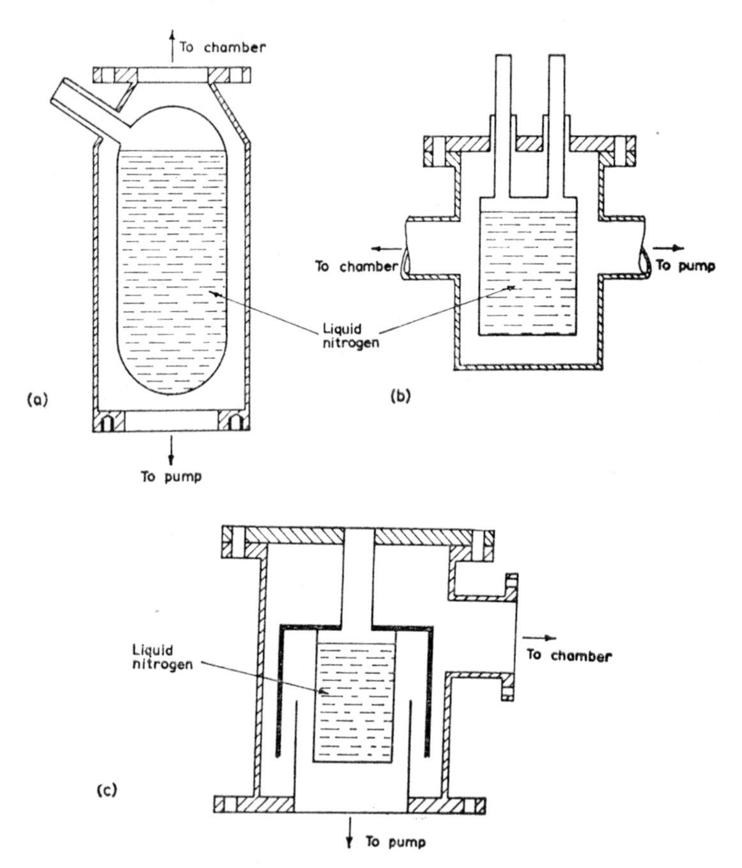

Fig. 23. Metal cold traps.

Post [22], to minimize the probability that pump oil vapour or its decomposition products will enter the chamber by migrating along the outer region of the trap, where the temperature is higher than at the central liquid-nitrogen container. To reduce loss of the liquid nitrogen by evaporation and to ensure trap surfaces which are not difficult to clean, the outer surface of the reservoir and the inner surface of the housing should both be highly polished. Roughly, the rate of loss of liquid nitrogen from such a cold trap, of outside diameter 10 cm, is 100 cu cm per hour.

Solid carbon dioxide ('dry ice') is a cheap coolant which is crushed into small fragments and made into a sludge with acetone or trichlorethylene or methyl cellosolve. It provides a minimum temperature in the trap of $-78°C$, at which the saturated vapour pressure of mercury is 10^{-8} torr, and that of water is 5×10^{-4} torr. It is not recommended except for small, glass, laboratory systems where an ultimate pressure of 10^{-6} torr is adequate, and should not be used in ultra-high vacuum systems.

Liquid air ($-183°C$) and liquid nitrogen ($-196°C$) are both excellent for cold traps, but so-called liquid air is chiefly liquid oxygen and hence supports combustion and is capable of violent explosive reaction with oils. Liquid air should therefore not be used in a glass cold trap above a glass oil diffusion pump, but is safe enough if a mercury pump is employed (presuming that the liquid air is not allowed to reach the oil in the mechanical backing pump!). Provided reasonable precautions are adopted, however, liquid air can be used in a metal cold trap above a metal vapour pump, though liquid nitrogen (often not so readily available) is undoubtedly preferable.

A problem with cold traps is that mercury or oil from the vapour pump may reach the upper part of the trap. Even a tiny speck of mercury at the top of a glass trap of the design shown in Fig. 22(*c*) will prevent ultra-high vacua being attained, as this section of the trap is clearly at a considerably higher temperature than $-196°C$, having a liquid-nitrogen filling. Furthermore, and especially in systems intended to attain pressures below 10^{-7} torr, it is important that the level of the liquid coolant is not allowed to fall unduly, leaving the upper parts of the trap ineffectively refrigerated. This is a difficulty with plant which is kept under vacuum overnight. Even though the trap can be isolated from the chamber by a vacuum valve, dropping of the coolant level, or its complete evaporation, will leave trap surfaces which release some of their previously condensed vapours. The remedy is to provide a trap of sufficient capacity so that it does not lose more than half its coolant over a period of about 15 hour. The metal traps shown in Fig. 23 are satisfactory in this respect, in sizes above 12 inch length and 4 inch outside diameter. The best practice, however, is undoubtedly to use a **liquid-nitrogen-leveller device,** the purpose of which is to maintain the level of the liquid gas in the trap above a certain minimum when the vacuum system is left standing over long periods. For good, standard, ultra-high vacuum work, such a provision is a necessity to ensure continuous day-to-day use without trap problems.

Many designs of leveller have been described in the literature; several of them do not really work satisfactorily! Commercial designs are also available. A useful one which has been found reliable and

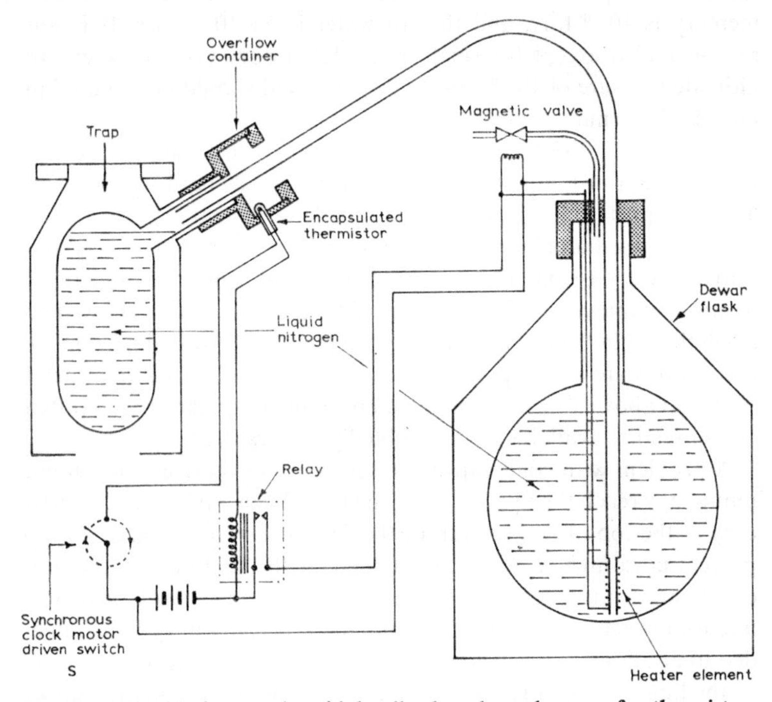

Fig. 24. A liquid-nitrogen (or air) leveller based on the use of a thermistor. (When switch S is closed, the relay and magnetic valve are closed, heater is on and liquid nitrogen is delivered to the trap. On overflow from the trap of the liquid nitrogen, the thermistor cools, its resistance increases and the relay is de-energized, so the magnetic valve opens, the heater is switched off, and the liquid nitrogen flow ceases. This sequence is repeated at a subsequent time decided by the period of revolution (e.g. 60 min) of the clock motor.)

cheap to construct is based on the use of a thermistor in conjunction with an electric heater, in a Dewar flask containing the liquid gas, and an electromagnetically-operated valve [23]. Its operation is explained in Fig. 24.

A cold trap will act as a pump for condensable vapours. With a filling of liquid air or liquid nitrogen, the chief action in this respect in most vacuum systems is that water vapour is pumped.

1.12. *The Molecular Drag Pump*

In 1912, Gaede [24] introduced the molecular drag pump, which operated relative to a backing pressure of the order of 1 torr, obtained by a mechanical pump, and which provided at its intake port a pressure of 10^{-6} torr and below. The action of this pump relies upon moving a metal surface very rapidly adjacent to a stationary surface. Within a sufficiently narrow gap between the two surfaces, the gas molecules present acquire momentum in the direction of the moving surface, because, on impact, they receive significant directed components of velocity. The dragging force exerted on the gas molecules establishes a pressure gradient and pumping action. However, for effective operation, the speed of the moving surface (obtained as that of a high-speed rotor) has to be very high to give a significant one-way motion to gas molecules having kinetic speeds of the order of 500 metre per sec at room temperature. The separation between the moving surface and the stationary one has to be very small because the equation applicable at viscous pressures is

$$p_1 - p_2 = 6l\omega\eta/d^2 \tag{1.14}$$

and at molecular pressures it is

$$p_1/p_2 = e^{b\omega} \tag{1.15}$$

where p_1 is the backing pressure, p_2 the intake (high vacuum) pressure; l is the length of the path between the two pressure regions; ω is the angular velocity of the rotor; d is the separation between the rotor surface and the stationary one (i.e. the stator surface); η is the coefficient of viscosity of the gas; and b is a constant, which decreases with the molecular weight of the gas and increases with decrease of d.

A satisfactory pumping speed can therefore only be obtained with rotor speeds of 5,000 r.p.m. or more, and with very small values of d.

In 1923, Holweck [25] designed an improved version of Gaede's pump, in which the rapidly moving member was in the form of an induction-motor-driven duralumin cylinder rotating at about 5,000 r.p.m. inside a bronze stator provided with spiral groves. The clearance between the rotor and stator had to be 0·02 to 0·05 mm though the spiral grooves had depths increasing from 0·5 mm at the backing pressure region of the pump to 5 mm at the high vacuum intake region. Siegbahn [26], in 1943, introduced a molecular drag pump in which the high-speed cylindrical rotor of Holweck was replaced by a disc driven at approximately 10,000 r.p.m. within a narrow cylindrical casing. With a disc diameter of 54 cm, a speed of 60 to 80 litre per sec for air was achieved at 8,000 r.p.m., at intake pressures of 10^{-2} to 10^{-6} torr and discharge to a backing pressure of about 1 torr.

The molecular drag pump has the great advantage of freedom from a pump fluid, and hence no back-streaming; it is consequently able to provide a 'clean' vacuum without the use of a cold trap. However, though the Holweck and Siegbahn types were used to some extent up to 1950, they did not prove popular because of mechanical problems associated with the high-speed drive, and because the small rotor-stator clearances demanded rendered the rotor liable to seizure on non-uniform thermal expansion, entry of small foreign solid particles, and even on a sudden inlet of gas.

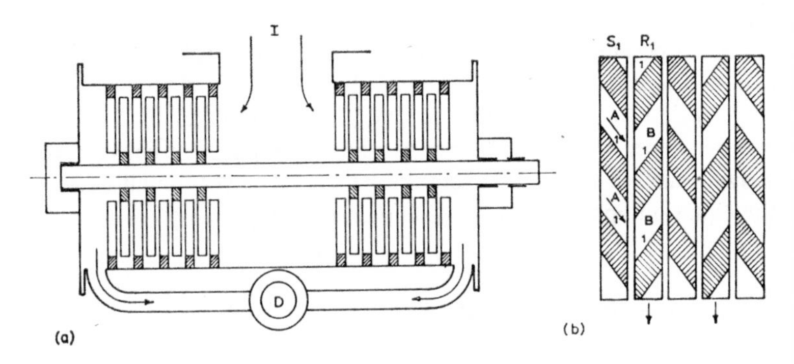

Fig. 25. Becker's molecular drag pump.

Since 1958, however, interest in the molecular drag pump has been revived with the advent of Becker's pump [27], sometimes called a **turbo-molecular pump**, which was made in Germany by Pfeiffer G.m.b.H. of Wetzlar. This new design avoids the problems of rotor seizure, because rotor-stator clearances of 1 mm are used; the rotor drive can then be at about 16,000 r.p.m.

The operation of Becker's pump is indicated schematically by Fig. 25(*a*). Like Holweck's pump, the housing is a metal cylinder, but otherwise it is considerably different. The rotor is in the form of two sets of parallel, separated, slotted discs, with twenty discs in each set (Fig. 26). This is driven from the axial ball-bearing-mounted shaft, so that these slotted spinning rotor discs are interleaved between similar stationary discs attached to the cylindrical pump housing.

The intake port (high vacuum side) of the pump is at I, with direct access to the region between the two sets of discs. The entering gas divides symmetrically into two streams, one to each of the two sets of rotor-stator discs. The discharge of the gas from the ends of the disc assemblies (left- and right-hand sides of Fig. 25*a*) is through a pair of pipes which join at D, the discharge outlet port, to which the mechanical backing pump is connected.

To provide ready passage through the disc assemblies and to impart momentum to the gas molecules towards the backing region, both the

Fig. 26. The rotor of Becker's molecular drag pump.

rotor and the stator discs are provided with a large number of radial slots (Fig. 26). The arrangement of these angled slots is shown schematically in Fig. 25(*b*). A rotor disc has everywhere an adjacent stator disc. The slits in the stator discs are inclined in the opposite way to those in the adjacent rotor discs, i.e. the inclined slots in the stator discs are mirror images of those in the rotor discs. Consider a slot A in the first stator disc S_1. The wall, 1, of this slot forms a wedge-shaped channel with the surface of the adjacent rotor disc, R_1. As R_1 moves in the direction of the arrow, gas molecules are driven in this same direction. The wall 1 in slot B of rotor disc R_1 also forms a wedge-shaped channel with the surface of the stator disc S_1, again resulting in gas being driven in the same direction as before. This action takes place similarly in the rest of the disc assembly. With discs a few mm thick, only a short channel is provided between discs, and consequently only small pressure differences are attained. However, with several discs, each containing many radial slots, all acting additively, a large total pressure difference between the intake and discharge ports is established. With a small pressure difference across only one pair of discs, the separation between neighbouring discs (compare the separation *d*, in equations (1.14) and (1.15), where $p_1 - p_2$ is small) can be 1 mm or more, without impairing the pump performance. The radial clearance between the rotor and the housing can also be 1 mm or more. The problems of rotor seizure encountered in previous molecular drag pumps with their much smaller clearances are therefore largely eliminated. If the angle between the slot and the surface of the disc is decreased, a higher ratio of intake to discharge pressure is achieved, but the pumping speed is smaller. Hence, the discs near the centre of the pump, where the high vacuum intake port is situated, have large slot-to-surface angles to ensure high pumping speed. Whereas the outer discs, near the backing pressure region, have slots at smaller angles to cope with the larger pressure ratios prevailing at the higher backing pressure.

Becker gives results for a pump with a rotor diameter of 17 cm and overall length of 65 cm, driven at 16,000 r.p.m. by an external a.c. motor of 0·3 kW in a system built of welded steel with metal gaskets and seals. The pumping speed for air is constant below 10^{-2} torr at 500 cu metre per hour (140 litre per sec). With a backing pressure of 10^{-1} torr, an intake pressure of 10^{-8} torr is achieved; if the backing pressure is reduced to 5×10^{-3} torr, an ultimate pressure of 5×10^{-10} is attained (suitable bake-out is presumably practised), the residual gas being hydrogen. The compression ratio (pressure at discharge/pressure at intake) provided by this pump is $1{\cdot}5 \times 10^{7}$ for air, but only 250 for hydrogen due to the smaller molecular weight of hydrogen (see equation 1.15).

The low compression ratio for hydrogen means that hydrogen will back-diffuse in this pump, so forming the chief contribution to the ultimate pressure at about 10^{-9} torr, despite the fact that the pumping speed for hydrogen is some 20% greater than for air because the light hydrogen molecules acquire greater speeds for a given solid surface impact. Sorption and desorption of gases from the pump and chamber walls play an important role in the operation of this Becker pump. In particular, the inlet of moist air should be avoided as far as possible, as it is likely to prolong, by a factor of about 10, the pumping time down to 10^{-6} torr.

The Becker molecular drag pump is promising (though rather expensive) in application to the problems of creating a 'clean' vacuum in such applications as pumping particle accelerators (especially as an intermediary between a getter-ion pump and a mechanical pump), large electron tubes, and mass spectrometers. But it is important to realize that oil from the mechanical backing pump can still reach the chamber unless an oil-vapour trap is included between the backing pump and the molecular pump.

1.13. *The Roots Pump*

The Roots pump used for the production of vacua is a comparatively recent development of a principle which has been applied to blowers for many years. It is a particularly useful pump for handling large quantities of gas in the pressure region between 10^{-3} and 10 torr, and so rivals the vapour ejector pump (section 1.7), though its capital cost is higher. It is backed by an oil-sealed mechanical rotary pump, and is of real value in large-scale engineering vacuum plants rather than laboratory-size systems.

In the Roots pump (Fig. 27), two figure-of-eight-shaped rotors or impellers mounted on parallel drive axles intermesh, with a small minimum clearance between them of about 0·1 mm. These rotors are driven in opposite directions at about 2,000 to 4,000 r.p.m. within a close-fitting housing; the drives to the impellers are synchronized by the use of timing gears. The gears and bearings are in water-cooled housings outside the main stator casing of the pump, with driving shafts to the rotors through seals in the stator end cover and with an external electric motor.

Consider that point in the rotation of the impellers where one impeller, R_1, is vertical and the other horizontal as shown in Fig. 27(*a*). With the directions of rotation as shown by the arrows, i.e. R_1 anticlockwise and R_2 clockwise, the intake port (high vacuum side) of the pump is to the right at I, and the discharge is to the left at D, where the backing pump is connected. After 90° of rotation, R_1 is horizontal and R_2 vertical (Fig. 27*b*), and some of the gas which was within the shaded volume of Fig. 27(*a*) connected to the intake port (and so the chamber) becomes isolated within the cross-hatched region of Fig. 27(*b*), between the pump housing and the impeller R_1.

This temporarily trapped gas cannot easily escape because the minimum clearance (at points 1, 2, and 3) between the impeller surfaces and the housing is only about 0·2 mm. Further, the discharge backing

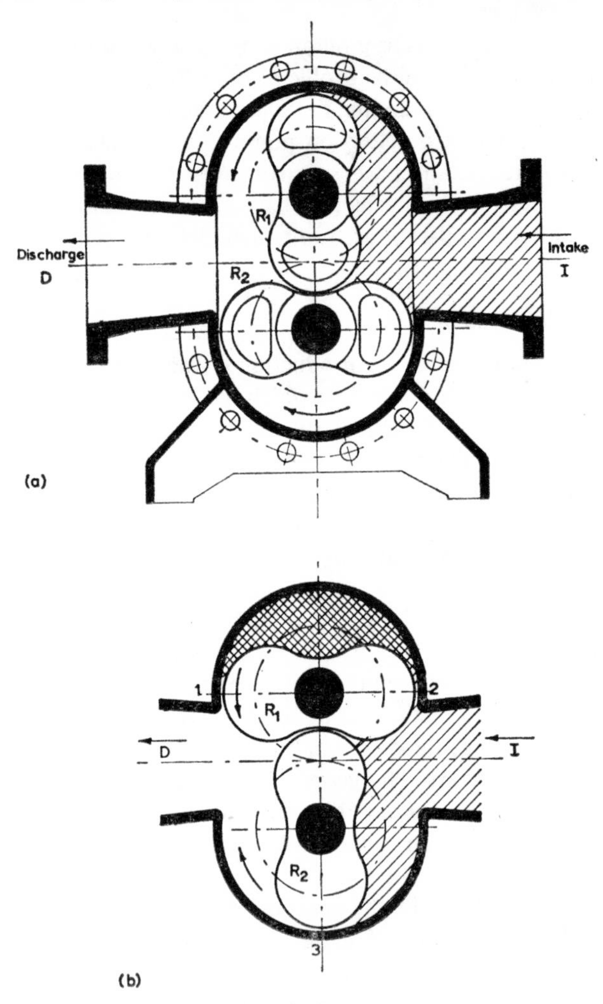

Fig. 27. The Roots pump.

region is isolated from the intake region by the 'slot' of about 0·1 mm width between R_1 and R_2. On further rotation of R_1 and R_2, this temporarily trapped gas is expelled to the discharge outlet; there it is compressed, because of the higher backing region pressure prevailing, and conveyed to the atmosphere by the mechanical backing pump. This process will occur with both impellers in turn; two 'samples' of the gas of volume represented by the cross-hatched region of Fig. 27(*b*) being discharged for each complete revolution of one of the impellers.

The impellers can be rotated at much higher speeds than in an oil-sealed mechanical pump because they do not touch one another: 2,000 r.p.m. is generally used for larger Roots pumps and up to 4,000 for smaller ones. Hence, very large pumping speeds can be obtained, though there will be back-diffusion of gas from the backing to the high vacuum side because of the finite gap clearances between the impellers. This backward streaming of gas will increase the higher the pressure ratio p_D/p_I (where p_D is the pressure at the discharge to the backing pump and p_I is the intake pressure to the Roots pump). However, there is an advantage in that the impedance to gas flow afforded by the clearance gap between the impellers will increase as the pressure p_I decreases, because then the m.f.p. of the gas molecules becomes longer and intermolecular collisions become insignificant. With discharge to the atmosphere – which is possible – the compression ratio is thus limited to 3 or 4; with a backing pressure of 1 torr, the compression ratio can readily be 10; whilst in the region of 10^{-2} torr, it can be more than 50.

The compression of the gas in a Roots pump will cause heating; it is not easy to convey this heat away and there is no oil-filling to assist. Excessive temperature rise could cause thermal expansion of the impellers, resulting in seizure. Clearly, the handling of large quantities of gas at high intake and discharge pressures would result in greater heat development; indeed, for most purposes the maximum difference of pressure across the Roots pump is limited to 50 torr. To avoid over-heating, some larger models have water-cooled plates in the discharge outlet, and others have cooling oil circulated through hollow axles to the impellers. But the usual procedure to avoid over-heating, and to avoid the use of an excessively large motor drive to the Roots pump, is to reduce the amount of gas compressed, by not bringing the Roots pump into full operation until the intake pressure has been reduced to 10 torr or less by the mechanical backing pump.

To do this, one of the following methods is usually employed.

(*a*) Set up the Roots pump and the oil-sealed mechanical backing pump as in Fig. 28, with the vacuum valve V initially open and the Roots pump idle. Evacuate the system to 1 to 10 torr by the backing pump alone, close valve V, and switch on the Roots pump.

(*b*) Control the action of the Roots pump motor by a pressure-actuated switch, so that the mechanical pump operates alone until the pressure is below, say, 1 torr. Meanwhile, during the pressure reduction from 760 to 1 torr, either the Roots pump motor is off and the impellers simply 'windmill' in the gas stream or the Roots pump is by-passed by a tube containing a pressure-monitored vacuum valve. When the chamber pressures decrease to 1 torr, the Roots pump motor is switched on and, if used, the by-pass valve is closed.

(*c*) The Roots pump drive is through a torque limiting device, such as a hydraulic torque converter or a fluid flywheel. Both pumps are switched on together, but the speed of the Roots pump is initially

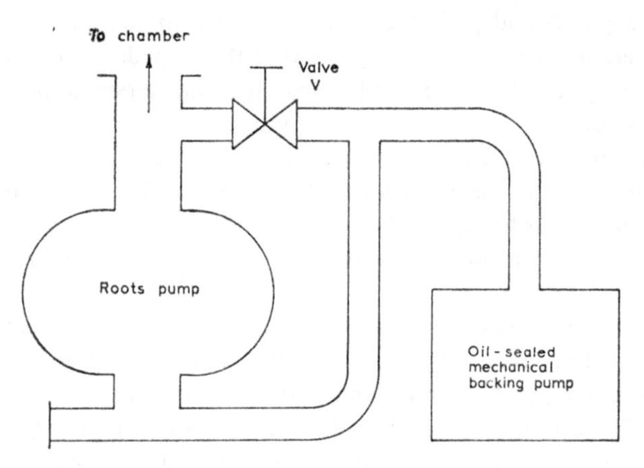

Fig. 28. Operation of a Roots pump with an oil-sealed mechanical backing pump.

limited to a maximum safe torque until, as the pressure drops, full drive is attained. This practice is adopted in the Roots pumps (mechanical booster pumps) marketed by Edwards High Vacuum Ltd.

(*d*) Balzers (Liechtenstein) Roots pumps incorporate an electrical contact controlled by thermal expansion of the rotating impellers relative to the casing; if thermal overload is approached, this contact trips the motor switch, so shutting down the pump.

The Roots pump will not suffer back-streaming difficulties like the oil vapour pumps, though there may be leakage of oil through the seal around the shaft to the external timing gear mechanism. The mechanical backing pump oil vapour can, however, enter the chamber via the Roots pump, though its effective vapour pressure will be noticeably reduced compared with using the oil-sealed mechanical pump alone. An oil vapour trap between the Roots pump and the mechanical pump is therefore advantageous.

Roots pumps are often used in series, one backing the other in a two-stage arrangement able to provide an overall compression ratio of 100 or more. Two-stage Roots pumps with a gas-ballast oil-sealed mechanical pump providing a backing pressure of 10^{-2} torr can attain an ultimate intake pressure of 10^{-4} torr or below, this ultimate being primarily due to oil vapour pressure.

The pumping of a certain mass of gas per sec corresponds to the throughput of a certain amount expressed as a pressure-volume product. Assuming a usual compression ratio in a Roots pump of 10, it is clear that the speed at the backing pressure of the mechanical backing pump need only be $\frac{1}{10}$ of the speed of the Roots pump at its intake pressure. Thus, the handling of large quantities of gas, particularly in the pressure range from 10^{-2} to 1 torr, is possible using the Roots pump with a mechanical pump of about $\frac{1}{10}$ the capacity that would be needed if it were used alone. In this respect, a Roots pump is often of advantage in large installations as an intermediary between a diffusion pump and a mechanical backing pump, because a comparatively small mechanical pump can be used.

Roots pumps are available from most vacuum equipment manufacturers, in sizes ranging from those with speeds of 20 litre per sec up to large installations having speeds of 23,500 litre per sec (50,000 cu foot per min). Typical intermediate size models are the R 150 and R 152 of Heraeus G.m.b.H. of Hanau, Germany: both have a speed of 41 litre per sec. The single-stage type, R 150, is able to provide an ultimate pressure of 10^{-4} torr, whilst the two-stage R 152 achieves 5×10^{-6} torr. In both cases, the maximum intake pressure tolerable is 20 torr. Fig. 29 shows their pumping speed against intake pressure characteristics when operated with a mechanical backing pump having a free air displacement of 420 litre per min.

Roots pumps are particularly useful in large installations for vacuum furnaces, impregnation plant, vacuum stills and concentrators, vacuum drying, and evacuating space simulators. In comparison with their chief rival, the vapour ejector pump, they are more expensive, and liable to damage on the entry of foreign solid particles, but are more compact, give less difficulty from oil vapour entering the chamber, and are less affected by sudden in-rushes of gas. In general, the Roots pump is a good choice for pumping large quantities of gas in the pressure range from 10^{-1} to 10 torr. Between 10^{-1} and 10^{-4} torr, the vapour ejector pump is preferred, whilst above 10 torr, the oil-sealed mechanical pump is usually best. The last pump is also usually preferable to the Roots pump if the required speed does not exceed about 100 litre per sec or 200 cu foot per min approximately.

1.14. *Getter Pumps and Getter-ion Pumps*

When molecules of a gas or vapour impinge on a solid surface *in vacuo*, they may be sorbed, i.e. retained by the solid material. **Sorption** is of two main types: **adsorption**, in which the molecules are re-

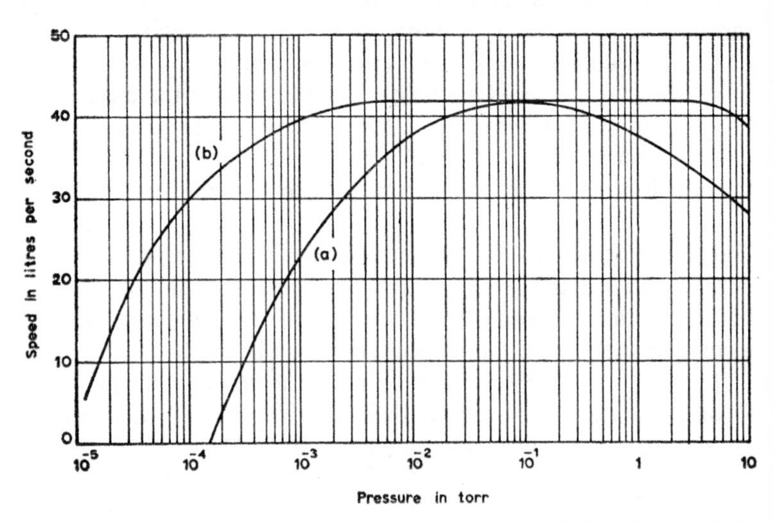

Fig. 29. Pumping speed against intake pressure characteristics of: (*a*) a single-stage; and (*b*) a two-stage Roots pump.

tained at the surface of the solid; and **absorption**, in which the molecules enter into the interior structure of the solid by a process of diffusion and/or solution. The classification into physical and chemical sorption is also important. In the former, the binding forces between the molecules of the gas and the solid are of the Van der Waal's type or of a polar nature, and are comparatively weak. In **chemisorption**, the incident gas combines with the surface atoms or molecules of the solid to form a chemical compound, and the binding forces are comparatively strong. The opposite process to sorption, known as **desorption**, is the release of gas by a solid. It is sometimes difficult in practice to distinguish between adsorption and absorption, and even between physical sorption and chemisorption. Physical sorption is increased by lowering the temperature of the solid surface: it is a process akin to condensation of the molecules on the surface. Chemisorption, on the other hand, is in general increased by raising the temperature of the solid.

The sorption of a gas by a solid is usually enchanced if the gas is ionized, and especially if the ions are accelerated. The sorption of oxygen by a reactive metal with which it forms oxides is little affected by ionization. However, physical adsorption, cases where diffusion of the gas into the metal occurs (e.g. with hydrogen), several cases of chemisorption, and, most important of all, the sorption of noble gases are significantly affected by ionizing the gas.

Consider, for example, the inert noble gas neon. At a temperature T, this will have a most probable kinetic energy of kT, where k is Boltzmann's constant equal to $1{\cdot}38 \times 10^{-16}$ erg per degC or 1/11,600 eV per degC (since 1 electron-volt (eV)$=1{\cdot}6 \times 10^{-12}$ erg). The neon atom (or any other gas molecule or atom) will consequently have a kinetic energy at 300°K (27°C) of 300/11,600 eV, i.e. 0·026 eV.

If the neon is ionized, the pair of ions formed consists of a positive ion, Ne^+, and an electron (negative ion formation is not possible in the inert gases, as the probability of an electron becoming attached to a neutral inert gas atom is zero). The positive ion formed no longer has an outermost closed electron shell, but a shell with one electron missing: it now readily acquires an electron to become neutral. On approaching a negatively-charged electrode (cathode) in a neon discharge, the positive neon ion will become neutralized in the immediate vicinity of the electrode by removing an electron from the electrode surface (an action demanding the work function energy of a few eV). If the cathode is at a negative potential of V with respect to the region where the positive ion was first formed, the neutralized positive ion, i.e. atom, will enter the electrode with an energy of VeV. With values of V of several hundred or thousand volt, it is clear that the atom can penetrate much more readily into the lattice structure of the metal with its energy of, say, 1,000 eV instead of the kinetic energy at room temperatures of only 0·026 eV. The extent of this penetration will depend on the size of the ion compared with the lattice spacing, and is thus more significant with the smaller atoms such as helium.

If N molecules or ions of a gas or vapour impinge on 1 sq cm of a solid surface per sec, and αN of these molecules are sorbed, α is defined as the **sticking coefficient**; its value will depend on the nature and energy of the incident molecules or ions and on the nature and temperature of the solid surface. The maximum possible value of α is clearly unity. In the case of the noble gas atoms impinging on a metal surface at room temperature, it will be vanishingly small; if such gas atoms are ionized and accelerated in an electric field, α can become appreciable.

A **getter** is a material (usually a metal) which is included in a vacuum system or chamber and reduces the gas pressure because molecules which reach it are sorbed. Once sorbed, it is vital, if the low pressure obtained is to be maintained, that subsequent desorption be very small. The binding forces between the sorbed molecules and the getter therefore need to be large, so, in general chemisorption at the surface or the formation of a bulk chemical compound is the more important requirement. Further, the metal-gas combination formed must clearly have a low vapour pressure.

To ensure a high rate of gettering, i.e. pumping speed for the gas by the getter, the requirements are a large effective area of the getter and a high sticking coefficient. The latter is normally only ensured by providing fresh films of the getter. The usual procedure is therefore to heat the getter metal in the residual gas in a chamber at a pressure not exceeding 10^{-2} torr (and usually below 10^{-3} torr), so that it evaporates directly into the vapour state and a fresh film of the metal is deposited on a nearby cold surface. The object is then that the pressure drops quickly below the initial pressure because of gas sorption by the getter. In this gettering action, two processes occur: **dispersal gettering**, in which the gas is taken up whilst the getter is being volatilized, i.e. whilst it is in the vapour state; and **contact gettering**, in which the metal film deposited on a surface will sorb incident molecules.

As an alternative to producing the getter film by evaporation *in vacuo*, the technique of **cathodic sputtering** may be employed. At a gas pressure usually below 10^{-1} torr (p. 255), an electrical discharge is set up between two electrodes in the gas by the application of a potential difference, V, across them generally exceeding 1,000 volt. The positive ions created in the discharge impinge on the cathode with energies up to V eV and dislodge from the cathode atoms of its metal. These are then deposited on surfaces in the vicinity, including the anode; a getter film of the cathode metal is thus obtained.

A wide variety of metals can be volatilized *in vacuo* or sputtered in a discharge to give a thin film deposit (p. 262). Several act as contact getters in the bulk form when heated, but there are only a few metals which have useful gettering properties: barium, hafnium, molybdenum, tantalum, thorium, titanium, tungsten, uranium, and zirconium. Of these, only titanium and zirconium, but particularly the former, have been extensively used in the design of vacuum pumps, whilst barium is the most valuable getter in electron tube processing (p. 232). Barium has been employed in some prototype designs of vacuum pump, but these have not been made available commercially.

The metal titanium with its extraordinary affinity for gas has thus been the first choice in the development of getter pumps. Titanium is available as a vacuum-melted metal in the form of wire or sheet containing approximately 0·008% by weight of hydrogen (i.e. 1 gram of this metal is able to release $6{\cdot}5 \times 10^5$ litre of hydrogen at 10^{-6} torr) so it is essential to degas it by heating to 1,300°C *in vacuo* before using it as a getter. Titanium has a melting point of 1,660°C and a vapour pressure of 10^{-1} torr at 1,742°C. Its sorption properties for gases are summarized in Table 1.4, extracted from extensive data given for a variety of metals by Holland. [28].

It follows from equation (3.7) that the rate at which nitrogen would be pumped by 1 sq cm of a surface which retained all the molecules incident upon it (i.e. sticking coefficient $\alpha = 1$) would be 11·7 litre per sec. From the Table 1.4, the figure for nitrogen on a fresh film of titanium (1 sq cm in area) is 3·0 litre per sec, corresponding to a sticking factor of 3/11·7, i.e. 0·26.

Three main types of pump have been introduced utilizing the gettering action of titanium: two are **getter-ion pumps** and the third is a **sublimation** or **getter pump**. All of them operate relative to a backing pressure established by either an oil-sealed mechanical pump or a sorption pump (section 1.15).

There are two kinds of getter-ion pump: in the first, titanium is volatilized by heating it to a temperature of about 2,000°C in a backing vacuum of 10^{-2} torr or below, and the gas is ionized by electrons from a thermionic filament which are accelerated to a positive grid; in the second, titanium is sputtered from a cathode of this metal in a discharge in the gas (initially at a backing pressure of 10^{-2} torr or below) formed by applying a p.d. of 1,000 volt or more between the cathode and an anode. The first type is known as an **evapor-ion pump** (actually a trade name of the Consolidated Vacuum Corp., U.S.A.), and the second as a **cold-cathode getter-ion pump** or **sputter-ion pump** (VacIon pump is a trade name of Varian Associates, U.S.A.).

Large evapor-ion pumps of intake port diameter 12 inch or more,

TABLE 1.4

Sorption characteristics of titanium

	Gas	*Initial sorption rate at 20°C per sec per sq cm in litre or litre-torr*	*Temperature for continuous sorption*	*Sorptive capacity in litre-torr per mg*
Titanium in form of initially fresh film	O_2	Chemical reaction		
	H_2	Chemisorption		
	N_2	3·0 litre		$1{\cdot}9$ to $2{\cdot}5 \times 10^{-3}$ at 30 to 300°C
	CO_2	4·3 litre		$4{\cdot}3 \times 10^{-3}$ at 20°C
	CO	12·0 litre		$3{\cdot}4$ to $4{\cdot}2 \times 10^{-3}$ at 30 to 200°C
	SF_6	Chemisorption		
	C_2H_2	Chemisorption		
	C_2H_4, CH_4, CCl_2F_2, NH_3	Chemisorption		
Titanium in solid form	O_2	2×10^{-3} litre-torr at 800°C	> 650°C with linear law of oxidation	9×10^{-2} at 800°C
	H_2	Diffusion	20 to 400°C	
	N_2	8×10^{-5} litre-torr at 1000°C	> 700°C	$1{\cdot}6 \times 10^{-1}$ at 1000°C
	CO_2	8×10^{-4} litre-torr at 1100°C	> 700°C	5×10^{-2} at 1100°C
	H_2O	Diffusion	300 to 400°C	

having speeds of several hundred litre per sec, have been built, in which titanium wire is fed continuously onto a tungsten-tantalum alloy post heated by electron bombardment. Though these pumps have been used in proton-synchrotron vacuum systems (Gould and Dryden [29]), it would seem unlikely that they will survive in competition with the more convenient cold-cathode getter-ion pump.

However, the evapor-ion pump in smaller sizes is still used considerably. The type shown in Fig. 30 is essentially a development from the Bayard-Alpert hot-cathode ionization gauge (section 2.7). Two tungsten filaments, F and F_z, are arranged as shown outside the cylindrical grid G within the glass envelope. Along the axis of G is a

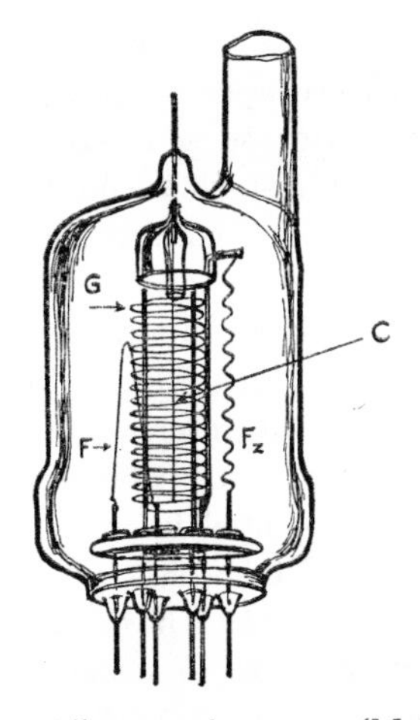

Fig. 30. A small evapor-ion pump (Mullard Ltd.).

thin wire ion collector electrode, C. The hairpin-shaped tungsten filament, F, is heated to about 2,500°C by the passage of electric current (6 amp). It emits electrons which are accelerated to the grid G, which is maintained at 200 volt (grid current $I_g = 10$ mA) with respect to F. The spiral tungsten filament, F_z, is loaded with zirconium wire coiled tightly round the tungsten. On heating F_z to about 2,000°C by the passage of electric current (maximum 18 amp), the zirconium evaporates and is deposited in the form of a thin film on the inside wall of the glass envelope. This film will getter the active gases and also, to a smaller extent, the inert gases, because of the ionization caused by electron impact. The central ion collector electrode, C, is at a potential of -100 volt with respect to F, the filament.

The pumping speed is approximately 0·3 litre per sec for air. To attain the ultimate pressure of 10^{-9} torr or below, of which these pumps are capable, the procedure is first to establish a pressure of about 10^{-6} torr by a vapour diffusion pump in the chamber, to which the evapor-ion pump is attached as an appendage. The chamber and pump are then baked until, on subsequent cooling, the pressure is 10^{-7} torr or below. The filament F is then heated with a current of 8 amp and a potential of about 300 volt is applied to grid G, so that the grid current is some 150 mA; the power dissipated in G by electron bombardment is then about 45 watt, sufficient to make it red hot. This heating is continued for 30 min; meanwhile, the filament F_z is heated by a current of 4 amp, to degas it. The evapor-ion pump is then allowed to cool. It and the chamber are isolated from the diffusion pump by closing a bakeable metal valve or by sealing-off at a constriction in a glass tube. The evapor-ion pump is run at recommended potentials and currents to cause getter-ion action and reduce the chamber pressure to the ultra-high vacuum region. The getter filament F_z can either be operated continuously at 5 to 6 amp or intermittently at 8 amp.

In the cold-cathode getter-ion pump or sputter-ion pump (Hall [30], Jepsen [31]), a titanium getter film is deposited by sputtering from a cathode of this metal. However, it would be of little use simply to set up plane-parallel electrodes of titanium in the residual gas at, say, 10^{-2} torr and apply a potential difference across these electrodes. Although a discharge would take place in the gas and the cathode would be sputtered by the incident positive ions formed in the discharge, it would not be possible to sustain adequate ionization at pressures below 10^{-4} torr, where the electron m.f.p. is about 300 cm (section 1.1), considerably greater than a convenient electrode separation. The cold-cathode getter-ion pump is therefore based on a **Penning discharge** (cf. the Penning gauge, section 2.6), in which the ionized gas between a pair of anodes and a cylindrical cathode between them is confined within a magnetic field. The essential arrangement of a single-cell pump is hence as shown in Fig. 31. The parallel disk-shaped plane cathodes, CC, have symmetrically between them a cylindrical anode, A, within a glass or stainless steel envelope. A uniform magnetic field of flux density, B, is established, with its lines of force along the common axis of symmetry of CC and A. The cathodes CC are of titanium sheet about 1 mm thick and the anode is also usually titanium but may be of non-magnetic alloy or copper. The p.d. between the anode and the common cathodes is about 2,000 volt and the magnetic flux density about 2,000 gauss.

The full mathematical theory of the Penning discharge is very complex, and for the getter-ion pump of this type it has not been fully worked out. However, simple equations indicate the important concepts. When the p.d. is established across the electrodes in the residual gas, a discharge will be set up and positive ions will travel towards the cathodes, whilst electrons will tend to move to the anode. The majority of the electrons can only reach the anode by having a component of velocity perpendicular to the axis of the system and so to the magnetic field lines. The maximum velocity which the electrons can acquire is v, given by

$$v = \sqrt{(2Ve/m)} \tag{1.16}$$

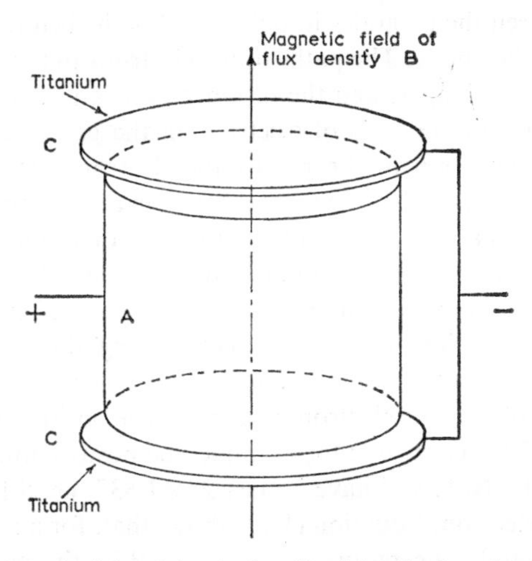

Fig. 31. The principle of the single-cell cold-cathode getter-ion pump.

where V is the anode-cathode p.d., e is the electron charge and m its mass.

Consider the extreme case where the electrons initially travel perpendicularly to the direction of the magnetic field lines with this velocity v. Within the flux density B, they will execute spiral paths with a radius R given by

$$R = mv/Be$$

Substituting for v from equation (1.16) gives

$$R = \sqrt{(2Vm/e)}/B \tag{1.17}$$

With $v=2{,}000$ volt$=2\times10^{11}$ e.m.u., $B=2{,}000$ gauss, and as $e/m=1{\cdot}76\times10^{7}$ e.m.u. per gram,

$$R = \sqrt{(4\times10^{11}/1{\cdot}76\times10^{7})}/2{,}000$$
$$= 0{\cdot}075 \text{ cm}$$

The majority of the electrons will not have such velocities perpendicular to the magnetic field lines, and hence will be confined to spirals of even smaller radii, because R decreases with the perpendicular velocity component. It is readily seen, therefore, that most of the electrons are confined to spiral paths having axes along or at small angles to the axis of symmetry of the electrodes. Consequently, they do not readily reach the anode, but oscillate backwards and forwards between the cathodes in tight spiral paths before finally being trapped by the anode. The paths of the electrons in the gas are therefore very much longer than the distance between the electrodes and compare with the m.f.p.'s of electrons in the gas even at very low pressures. Ionization of the gas by the electrons therefore persists down to pressures of 10^{-10} torr and below in a Penning discharge, as compared with an effective limit at a pressure of about 10^{-4} torr in a cold-cathode discharge not confined in a magnetic field.

Within a gaseous discharge, it is the electrons which cause ionization; the positive ions have little direct ionizing ability, but do cause electrons to be ejected from the cathode on impact, which then add to the number of ionizing electrons available. The positive ions are much heavier than the electrons; for example, the positive ion of the nitrogen molecule, N_2^+, will have a mass $28\times1{,}837$, i.e. $5{\cdot}14\times10^4$ times that of the electron. Equation (1.17) shows that, for a given accelerating potential V, the radius of curvature R of the particle path is proportional to $\sqrt{m}$, where m is the particle mass. The nitrogen ions will therefore travel along paths of which the maximum radii of curvature are $10^2\sqrt{5{\cdot}14}$ or about 164 times those of the electrons, that is about 12 cm. The positive ions moving near the axis will be more tightly spiralled than this, but it is seen from the geometry of the electrodes that they have no difficulty in reaching almost directly the cathodes at either end of the anode.

At the cathodes, the positive ions arriving with energies of V eV (for singly-charged ions), where V is 2,000 volt or more, will have the following actions: ejection of titanium atoms (sputtering); ejection of electrons (which assist the ionization); ejection of previously sorbed gas at the cathode surfaces (gas sputtering). Some of the positive ions will also become sorbed at the cathode and so assist in reducing the pressure in the device. This cathode sorption is important for the inert gases. But the chief cause of gas sorption of the predominant active gases is gettering by the sputtered titanium film which, within the almost enclosed electrode arrangement, will be chiefly deposited on the anode. Thus, the device acts as a pump where the gas is mostly finally sorbed at the anode in the form of a titanium compound (e.g. oxide for oxygen and nitride for nitrogen), whilst the inert gases are about 90% sorbed in the cathodes.

Small, single-cell, cold-cathode getter-ion pumps are available commercially with speeds of 1 to 2 litre per sec (e.g. Mullard VPP-2) and are used like the small evapor-ion pumps as appendages to vacuum chambers. The multi-cell pumps having an 'egg-box' anode between large titanium plate cathodes with a stainless steel casing (Fig. 32) have, however, brought this type of pump to considerable prominence in recent years. Initially, they were developed commercially by Varian Associates in America. Such pumps are now available in a wide range of sizes from a number of manufacturers. They have pumping speeds for air from 5 to 5,000 litre per sec, anode-cathode potentials of 2,000 to 10,000 volt, and magnetic flux densities usually about 2,000 gauss conveniently supplied by permanent magnet assemblies directly attached to the exterior of the pump casing. The pumping speed of the smaller designs is limited to some extent by the small conductance paths to the interior of the electrodes, but this is avoided in larger designs by incorporating the electrode cell units within the inner wall of an essentially cylindrical housing.

These pumps have the following advantages.

(*a*) There is no fluid as in a vapour diffusion pump, so back-streaming is not possible. Hence, cold traps and baffles are not necessary and the vacuum obtained is 'dry'.

(*b*) The pumping speed is essentially constant from 10^{-5} down to 10^{-8} torr (Fig. 32*c*). The ultimate pressure can be reduced to 10^{-10} torr provided the chamber and the pump are subjected to bake-out, and, of course, provided gaskets and other components used have insignificant outgassing after bake-out. The pump itself can be baked under vacuum to 400°C provided the magnets are removed; with the magnets in place, baking to 250°C is possible with some designs.

(*c*) The pressure attained for a given gas is directly proportional to the discharge current. Providing calibration is undertaken, the measurement of the current through the pump by a suitable meter in series with the power-pack can be used to measure the pressure; so

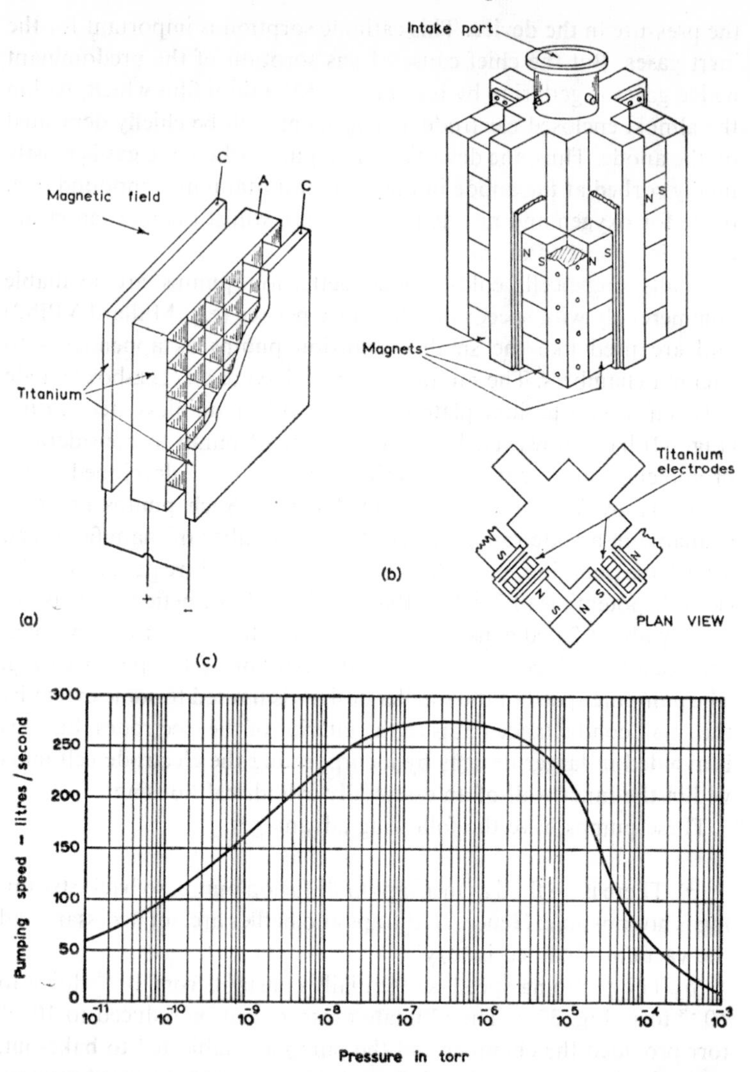

Fig. 32. (*a*) The multi-cell cold-cathode getter-ion pump. (*b*) Commercial cold-cathode getter-ion pump. (*c*) Typical pumping speed against pressure characteristic.

the pump can act as its own gauge. It must be stressed, however, that misleading results can be obtained by this practice because the pressure in the chamber to be evacuated may be different from that in the pump; also the calibration depends on the gases present. Monitoring of the pressure-dependent discharge current can also be used as the basis of a leak-detection device (section 3.11) when these pumps are used.

There are also the following disadvantages.

(*a*) The lifetime of the pump is proportional to the current in the discharge and the pressure attained. At a pressure of 10^{-6} torr, this lifetime is between 10,000 and 50,000 hour, but this depends very much on the previous history of the pump. The life is limited by flaking-off from the anode of sputtered titanium accompanied by the release of gas and even metallic particles which short out the electrodes, and also the erosion of the cathode plates by sputtering. The difficulty is that the lifetime will only be 10 to 50 hour at 10^{-3} torr, so it is essential to avoid unduly long pumping times in traversing the pressure region from the value provided by the backing pump down to 10^{-4} torr and below. In practice, this means that these pumps are not a satisfactory choice for vacuum plants in which considerable gas throughputs are handled, where the ultimate pressure is above 10^{-5} torr, or where processes are undertaken requiring repeated operation cycles in which the chamber is let down to atmospheric pressure two, three, or more times a day.

At higher pressures and discharge currents in the pump, the demands on the power-pack are considerable. A pump with a speed of 150 litre per sec for air will pass such a small current at 10^{-6} torr that the electrical load is only 10 watt, yet, at 10^{-4} torr or more, the power dissipation will be 1,000 watt and above, leading to overheating of the pump. Some models have water-cooled walls to cope with this, but the simple, usual practice is to ensure that the supply voltage falls off at higher load currents by providing a power-pack with sufficient internal resistance.

(*b*) They are expensive compared with vapour diffusion pumps of the same capacity.

(*c*) The operation is adversely affected severely by hydrocarbon vapours from an oil-sealed mechanical backing pump. To avoid this, either a cold trap is installed in the backing line between the getter-ion and the backing pump to condense out oil vapours, or, as is frequently the case, a sorption pump based on molecular sieve materials is used for backing (section 1.15).

(*d*) The discharge may be difficult to initiate at pressures below 10^{-7} torr, if the H.T. supply has been switched off temporarily. This is not generally a serious problem.

(*e*) A phenomenon known as **argon instability** is often exhibited by these pumps, if they have pumped considerable quantities of air or inert gases. The usual cause of this effect is pumping for several days against an air leak, which limits the pressure attained to 10^{-5} torr or above. The atmosphere contains 0·93% by volume of argon. The pumping speed for the inert gases will be some 100 times smaller than for nitrogen and oxygen, consequently, the residual atmosphere in the pump at low pressures becomes enriched with inert gases, particularly argon, which is nearly 500 times more plentiful in the atmosphere than all the remaining inert gases together. The argon will be mostly pumped to the cathodes where it will be physically adsorbed. As further positive ions of all the gases present arrive at the cathode, this adsorbed argon may be released by gas-sputtering. The effect shows up as a sudden gas pressure pulse to 10^{-4} torr or more, occurring fairly regularly at times separated by intervals of anywhere between several min and several hours, depending on the previous history of the pump.

Two ways of minimizing argon instability to a level where it is insignificant are: (*a*) to use a triode design of cold-cathode getter-ion pump (Fig. 33*a*) introduced by Brubaker [32]; and (*b*) to machine a series of channels in the cathodes of a diode design (Jepsen *et al.* [33]) (Fig. 33*b*). The second version is preferred because the power supply arrangements are less complex.

When an accelerated positive ion arrives at a cathode, the sputtering ratio (number of metal atoms released to number of incident positive ions per sq cm per sec) will increase with angle of incidence, i.e. be least for normal incidence (Wehner [34]). On the other hand, sorption of the ion by the cathode is greater the smaller the angle of incidence. In the triode pump, the perforated titanium cathodes have outside each of them an auxiliary plate electrode of titanium at a positive potential smaller than the anode potential. Sputtering of titanium will predominate at the perforated cathodes because the ions strike the walls of the perforations at large angles of incidence, whereas it will be less at the auxiliary electrodes because there incidence is normal and the positive ion energies are reduced by the positive potential. However, inert gas sorption will be at the auxiliary electrodes rather than at the cathodes, and these adsorbed gases are not only less likely to be released by arriving positive ions but are also covered over with sputtered titanium films from the cathodes.

In the diode design with channelled or slotted cathodes, the positive ions will be incident at glancing angles at the walls of these channels

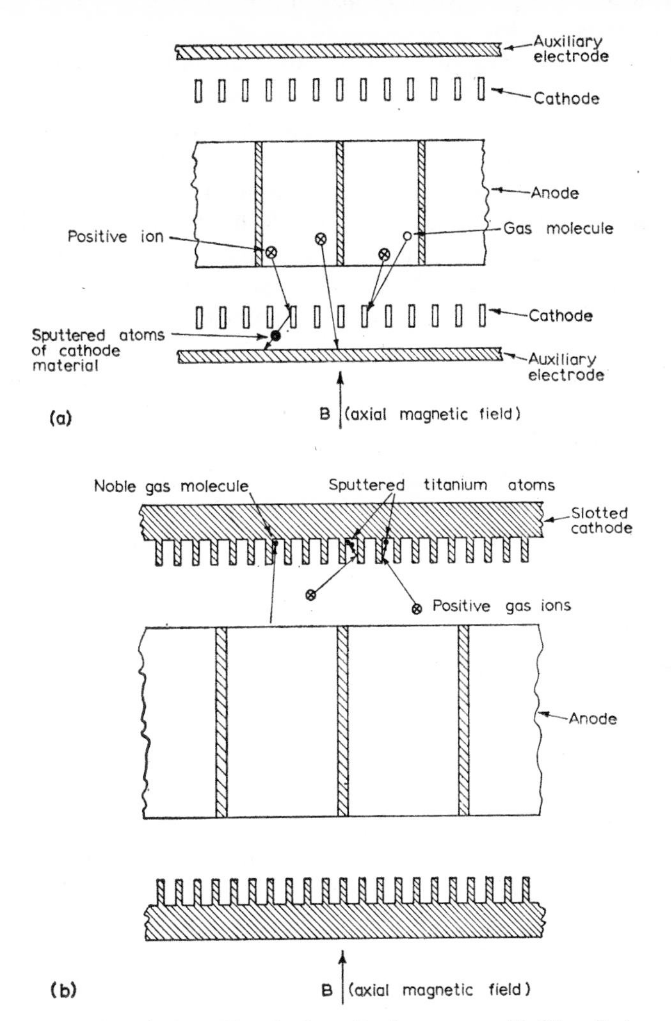

Fig. 33. (*a*) The triode cold-cathode getter-ion pump. (*b*) The diode pump with channelled cathodes.

but along the normals to their bases. Hence, sputtering will occur chiefly from the walls, whereas inert gas sorption will occur chiefly at the bases of the channels. At these bases, desorption on further positive ion incidence is less likely, and the sorbed gas will become plastered over with sputtered titanium films from the channel walls.

Recently, considerable interest has been shown in the **titanium**

sublimation pump or **getter pump**. This can provide very high speeds and forms a most promising and considerably more economical alternative to the evapor-ion pump with continuous titanium feed and its attendant difficulties. Clausing [35] has described a sublimation pump used to attain very high pumping speeds for hydrogen and its isotopes in thermonuclear research. An aluminium cylindrical tank of diameter 36 inch and height 36 inch contained a removable water-cooled copper liner which lasted for 150 days. This chamber is first evacuated by a 6 inch diffusion pump separated from the tank by a freon-cooled baffle at $-20°C$ and an isolation valve. The titanium wire, of diameter 0·035 inch, to be evaporated is coiled in two layers around the centre 7 inch length of a tantalum rod, of total length 10 inch and diameter 0·17 inch, with closely wound 0·03 inch diameter niobium wire between the titanium and the tantalum. The evaporation of the titanium was achieved by a current of about 500 amp through the loaded tantalum filament which had a p.d. of 2 to 3 volt across it. The niobium alloyed with the titanium on heating permitting faster evaporation. About 3 gram of titanium per hour could be evaporated with a filament life of 4 hour. A vacuum lock is provided at the top of the sublimation pump to permit replacing the loaded filament as required.

The titanium film deposited on the water-cooled walls will have a pumping speed per unit area, due to gettering action, depending on the sticking coefficient of the gas molecules to the film and the number of molecules incident per unit area per sec. Approximately, these speeds in litre per sec per sq cm for the various gases are: CO, 10·2; CO_2, 4·65; H_2, 3·1; N_2, 2·3; O_2, 1·7; and zero for the inert gases and methane; with the walls at room temperature, and at pressures below 3×10^{-7} torr. Above this pressure, the speed falls off rapidly. In Clausing's pump with a deposited film area of about 28,000 sq cm, the maximum speed obtained for hydrogen was about 80,000 litre per sec. This speed is within the pump chamber and would, of course, be considerably restricted if this pump were connected by a tube to a separate chamber. Without bake-out, the ultimate pressure obtained is about 3×10^{-8} torr.

The inert gases are not pumped by the gettering action of titanium; to ensure removal of these gases, the titanium sublimation pump is provided with an appendage in the form of a cold-cathode getter-ion pump. This addition combined with bake-out of the system enables ultimate pressures of 2×10^{-9} torr to be achieved (Holland and Harte [36]).

The titanium sublimation pump combined with the cold-cathode getter-ion pump is a valuable assembly for achieving very high pumping speeds at pressures below 10^{-6} torr, and is likely to find wide application in nuclear engineering plants and to the pumping of space simulator chambers. Very low pressures are achievable by condensing the titanium on liquid-nitrogen-cooled walls. The development of this system in conjunction with liquid-helium or liquid-hydrogen cryopumping (section 1.16) forms a promising approach to the problem of achieving ultra-high vacua in large chambers.

1.15. *Sorption Pumps*

A sorption pump (or trap) consists of a container with a filling of a renewable material, which has a large effective surface area and sorbs gas. Two materials useful for providing such a surface are activated charcoal and activated alumino-silicates, known as **molecular sieves.**

Sorption pumps based on one or other of these materials are useful alternatives in some circumstances to the oil-sealed mechanical pump, as a means of reducing the pressure in a chamber from atmospheric to about 3×10^{-3} torr, though this limit depends considerably on the construction and materials used. Their outstanding advantage over the mechanical pump is that there is no oil-filling and consequently no oil vapour. As a backing pump to the getter-ion pump, the performance of which is vitiated by the presence of oil, the sorption pump is therefore particularly attractive. It has led to considerable use of silent vacuum systems, with no moving parts, consisting of a cold-cathode pump with sorption pump. Indeed, in several laboratories, such systems are much preferred for obtaining ultra-high vacua instead of the use of oil diffusion pumps with cold traps (section 3.8).

Activated charcoal is prepared by subjecting coconut-shell charcoal to a heat treatment, and then baking it *in vacuo* to drive out occluded gases. The resulting granular material then has an effective surface area of about 2,500 sq metre per gram. It will absorb gases at room temperature, but this sorption is greatly increased by lowering the temperature to that of liquid nitrogen at −196°C.

A molecular sieve material is generally preferred to activated charcoal in the recent development of sorption pumps because its sorption capacity is greater at low temperatures. It also has more consistent properties, and, unlike charcoal, the pellets used do not become readily transferred undesirably into parts of the vacuum system away from the container. The chief disadvantage of molecular sieve materials compared with charcoal is their considerably smaller thermal conductivity: they are not so readily cooled throughout when in a container which is surrounded by liquid nitrogen in a Dewar flask.

The popular molecular sieve materials are types 4A, 5A, and 13A (British Drug Houses), having pore diameters of 4, 5, and 13 Ångstrom units (1 Å = 10^{-8} cm) respectively. Type 5A has been much used. It is a calcium alumino-silicate, available as conveniently handled pellets, the best being of the $\frac{1}{8}$ inch size. To provide effective pumping, this material is placed in a container which is immersed in a Dewar flask filled with liquid nitrogen or liquid air. Two main considerations affect the container design: the poor thermal conductivity in relation to attaining a low temperature throughout the pellets; and the fact that, to achieve satisfactory pumping speed, the containing

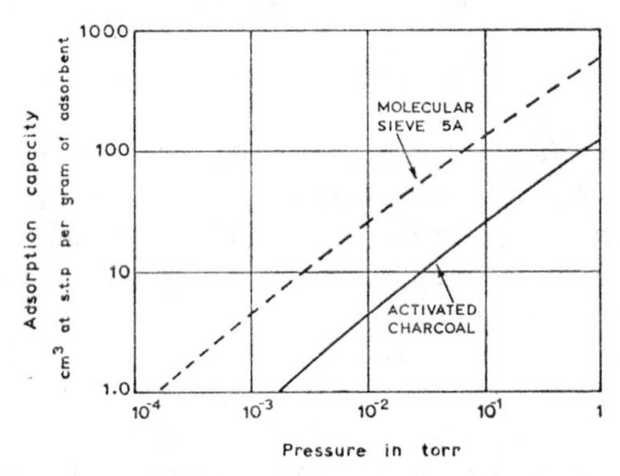

Fig. 34. Isotherms for adsorption of dry air by molecular sieve 5A and by activated charcoal at −196°C.

tube needs to have a wide bore to ensure ready flow of gas to the sorbent. These requirements are contradictory; the former demands a long narrow container and the latter a short wide one. Bannock [37] used a container for the molecular sieve pellets in the form of a tube of thin-walled copper-nickel alloy to ensure good heat transfer, but stainless steel is a good alternative, and glass may be used. This tube is best about 2 cm in diameter and 60 cm long. To increase pumping capacity, a number of such containers are used in parallel, each immersed in its own Dewar flask containing liquid nitrogen.

Adsorption isotherms for dry air, given by Bannock, compare molecular sieve 5A with activated charcoal, at −196°C (Fig. 34), and are approximately straight lines on log-log graph paper.

A typical, simple, molecular sieve sorption pump attached to a chamber and acting as a backing pump to a cold-cathode getter-ion pump is shown in Fig. 35 (*see* also section 3.8). This is a particularly useful set-up, introduced first by Varian Associates Ltd., as the getter-ion pump is kept free of the oil vapour which a mechanical pump would produce. (Mainwaring [38] describes the advantages of using an Edwards EZ60 molecular sieve pump in conjunction with a carbon-bladed oil-free rotary backing pump, to evacuate to 10^{-2} torr, large systems subsequently evacuated by a sputter-ion pump.) As a sorption pump can only adsorb a limited quantity of gas, after which it

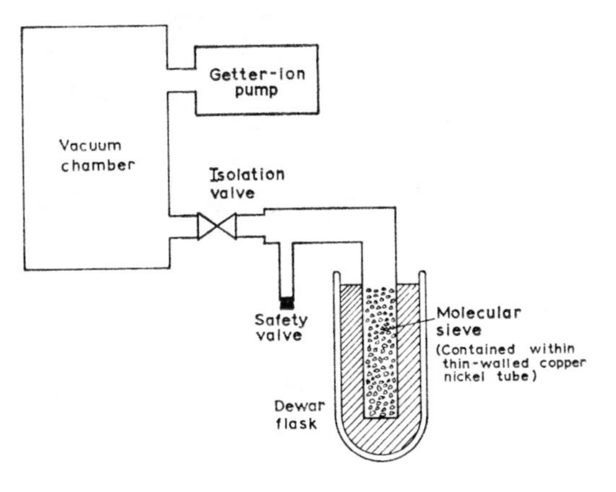

Fig. 35. Vacuum system comprising a molecular sieve sorption pump used to establish a backing pressure for a cold-cathode getter-ion pump.

will become saturated and begin to desorb, it is essential to provide an isolation valve between it and the getter-ion pump. This allows it to be periodically shut off, opened up to the atmosphere (after removing the liquid nitrogen!) then it desorbs the occluded gas, especially if baked, or the sieve material is replaced, if required. A safety valve (which can be simply a neoprene bung in a side tube to the sorption pump for backing pressures down to 10^{-2} torr) is essential, because dangerous over-pressures of gas can build-up in the isolated pump withdrawn from the liquid nitrogen.

Difficulties with the use of molecular sieve sorption pumps are:

(*i*) that argon and hydrogen are much less readily pumped than oxygen and nitrogen, so a cold-cathode getter-ion pump backed by one of these devices may exhibit its argon instability (section 1.14);

(*ii*) water vapour is preferentially sorbed and may achieve concentrations in the molecular sieve which inhibit prematurely its sorption of other gases.

To minimize this second possibility, the molecular sieve material is removed periodically and heated to 350°C (but not above 500°C) for 1 to 2 hour at atmospheric pressure, to remove water. It is then sealed from the atmosphere in a container, if stored before use.

Two molecular sieve pumps may be used in series to achieve pressures much less than 10^{-2} torr. In a glass system comprising two molecular sieve 5A traps in series separated by a greaseless glass stopcock, the cooling of one of these traps with liquid air produced a pressure of 2×10^{-2} torr, as recorded by a Pirani gauge in the second trap. The stopcock was then closed and the second trap immersed in liquid air. The pressure then decreased to 10^{-5} torr and below, as recorded by a hot-cathode ionization gauge, with a pumping speed for air at 10^{-4} torr of 0·02 litre per sec per gram of molecular sieve 5A.

1.16. *Cryopumps*

Cryopumping is a technique in which a cooled surface at a temperature T_c°K (T_c°K$=T_c$°C$+273$) is exposed to the gases in the chamber to be evacuated, on the basis that gas molecules impinging on this surface may be condensed there. The cold traps and sorption pumps already described are cryopumps in this sense, but generally the term is used to denote those devices in which T_c°K is so low that gases such as nitrogen and oxygen will be condensed.

In practice, it is not worthwhile to use cryopumping for evacuation of a chamber from atmospheric pressure but to pump first by conventional methods to about 10^{-3} torr. Apart from other considerations discussed below, pumping from 760 torr would result in excessive quantities of condensed gases and vapours on the cooled surface, which would introduce undesirable layers of poor thermal conductivity.

Assume that a gas of molecular weight M at a temperature T°K and molecular pressure p has a condensation coefficient of unity at a cooled surface of area A sq cm. From equation (1.9) it follows that the number of molecules of this gas which are condensed on the surface per sec is given by

$$N = \frac{p\mathrm{N}A}{\sqrt{(2\pi M\mathrm{R}T)}} \tag{1.18}$$

As N molecules occupy a specific volume V at a pressure p, where

$pV = \mathrm{R}T$, the volume occupied by N molecules at this pressure is $N\mathrm{R}T/p\mathrm{N}$. Defining S, the speed of pumping by the cold surface (known as the **cryopanel**), as the volume of gas condensed per sec at a pressure p, it follows that $S = (N\mathrm{R}TA)/(p\mathrm{N})$, and on substituting for N from equation (1.9)

$$S = A\sqrt{\left(\frac{\mathrm{R}T}{2\pi M}\right)} \tag{1.19}$$

Substituting $\mathrm{R} = 8{\cdot}314 \times 10^7$ erg per degC per mole and $T = 293°\mathrm{K}$ (assuming that the gas in the space above the cold surface is at room temperature, implying that the area of this surface is small compared with the surface of the inner walls of the chamber in which it is situated)

$$S = \frac{6{\cdot}25 \times 10^4 A}{\sqrt{M}}\ \mathrm{cm^3\ sec^{-1}}$$

$$= 62{\cdot}5A/\sqrt{M}\ \text{litre sec}^{-1} \tag{1.20}$$

A cold trap filled with liquid nitrogen, which has an effective surface area A at a sufficiently low temperature to condense water vapour, will therefore have a speed of pumping of water vapour given by substituting $M = 18$ (the molecular weight of H_2O) in equation (1.20)

$$\therefore\ S = \frac{62{\cdot}5}{\sqrt{18}} = 14{\cdot}7A\ \text{litre sec}^{-1}$$

If the condensation coefficient of the molecules at the cold surface is less than unity, the speed will be less than the value given by equation (1.20). As condensation coefficients are often not known with certainty, a preferable method of calculating the speed is to put p_1 as the partial pressure of the condensable gas at the temperature T of the chamber as a whole, and p_2 as the partial pressure of this ga at the cold surface at temperature T_c. The number of molecules arriving per sec at the cooled surface of area A is therefore N_1, given by equation (1.18) as

$$N_1 = \frac{p_1 \mathrm{N} A}{\sqrt{(2\pi M \mathrm{R} T)}}$$

corresponding, on the same reasoning as before, to an arrival pumping speed of

$$S_1 = \frac{N_1 \mathrm{R} T A}{p_1 \mathrm{N}} = A\sqrt{\left(\frac{\mathrm{R}T}{2\pi M}\right)}$$

measured at the pressure p_1.

The number of molecules leaving the cold surface is

$$N_2 = \frac{p_2 NA}{\sqrt{(2\pi MRT)}}$$

corresponding to a speed measured at the pressure p_1 (the pressure in the chamber) of

$$S_2 = \frac{p_2 A}{p_1} \sqrt{\left(\frac{RT}{2\pi M}\right)}$$

The net speed of pumping is therefore

$$S = S_1 - S_2 = A\sqrt{\left(\frac{RT}{2\pi M}\right)}\left[1 - \frac{p_2}{p_1}\right]$$

which, at $T = 293°K$ (20°C) and substituting for R, gives

$$S = \frac{62{\cdot}5A}{\sqrt{M}}(1 - p_2/p_1) \text{ litre sec}^{-1} \qquad (1.21)$$

Clearly, to achieve effective cryopumping by means of a cold surface, the vapour pressure p_2 of the condensed gas at the cold surface temperature T_c must be much less than p_1, the pressure of the gas in the chamber. The ultimate pressure attainable will be p_2 as measured at the temperature T_c. This value of p_2 is not usually simply the vapour pressure p_v of the condensed gas at the temperature of the cold surface, because the way in which the condensate is bonded to the surface will have an influence. Moreover, as regards the ultimate pressure as measured by a gauge within the cryopump, it must be remembered that the cryopanel usually forms only a small fraction of the internal area of the system. Therefore, gas molecules reaching the recording gauge are likely to have energies corresponding more to the temperature T of the system as a whole than to T_c. The ultimate pressure recorded will hence be decided chiefly by $p_2\sqrt{(T/T_c)}$.

The two most useful coolants for cryopumping by means of a cold surface are liquid hydrogen and liquid helium, of which the boiling points at atmospheric pressure are −252·8°C (20.35°K) and −268·9°C (4.25°K) respectively. Of these two, liquid helium is preferred but is considerably more costly. It has a latent heat of vaporization of only 650 cal per litre, as compared with 7,640 cal per litre for liquid hydrogen, and 48,000 cal per litre for liquid nitrogen. It is therefore most important in the design of a liquid-helium cryopump to prevent excessive evaporation by reducing thermal radiation and conduction to the liquid helium.

At the temperature of liquid hydrogen (20·35°K), the vapour pressures of all gases are in the ultra-high vacuum range with the exceptions of helium, neon, and hydrogen itself. For example, the vapour pressures at 20°K of some of the gases commonly encountered in vacuum systems are approximately: argon, 10^{-12} torr; nitrogen, 10^{-11} torr; oxygen, 10^{-13} torr; but neon has a vapour pressure of about 50 torr at this temperature. As neon forms 0·0018% of the atmosphere (partial pressure in air at 760 torr $= 1{\cdot}4 \times 10^{-2}$ torr),the use of liquid hydrogen is practical provided the bulk of the gas is first removed by conventional pumps before cryopumping is employed.

At the temperature of liquid helium (4°K approximately), all the gases will solidify with the exception of helium itself, but hydrogen will exert a vapour pressure of about 10^{-7} torr so that to produce ultra-high vacua additional means of removing hydrogen (e.g. evolved from stainless steel) may be necessary.

Helium is present in the atmosphere, to the extent of 0·0005%, equivalent to a partial pressure of $3{\cdot}8 \times 10^{-3}$ torr. Hence, a liquid-helium cryopumping surface used alone to pump a chamber from atmospheric pressure cannot attain an ultimate pressure below 4×10^{-3} torr approximately. Manifestly, it is therefore necessary to reduce the chamber pressure to, say, 10^{-4} torr before cryopumping, when the ultimate pressure due to helium gas remaining will be ideally only $4 \times 10^{-3}/760{,}000$, i.e. 5×10^{-10} torr approximately.

The attractive features of cryopumping with liquid helium or liquid hydrogen are the ability to create pressures in the ultra-high vacuum range, and that a very high pumping speed is possible, with the cooled surface either placed directly in the chamber or in a readily accessible side chamber, so that little restriction to gas flow is obtained.

To minimize radiation of heat to the cold surface – especially necessary with liquid helium because of its low heat of vaporization – shielding cooled with liquid nitrogen (–196°C and with much large vaporization heat) is interposed between the cold surface and the chamber walls and other surfaces at high temperatures. In this design, the important factors are: (*i*) the need for good liquid-nitrogen circulation to a shield of a metal of high thermal conductivity; (*b*) the shielding arrangement should not impede gas flow to the cryopumping cold surface, and yet should have a thermal radiation transmission from the surface at room temperature to the cold surface of 2% or less; and (*c*) heated sources such as gauge filaments must be carefully shielded from the cold surface. These shields are usually polished for low emissivity except for surfaces which might reflect heat from the vacuum chamber onto the cryopanel. They also serve to precool the gas by arranging that the gas molecules make several collisions with the radiation shields before reaching the cryopanel. They also pump water vapour.

Connecting tubes from the liquid-helium (or hydrogen) tank to the cold, cryopumping surface must be thermally insulated and kept as short as possible. To conserve helium, either a closed-cycle refrigeration system is used on a large plant, or the helium emerging as gas

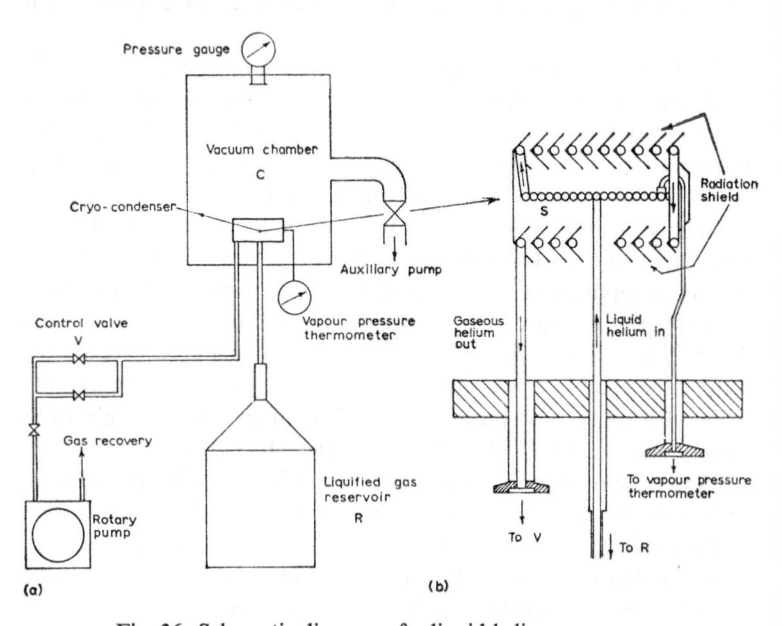

Fig. 36. Schematic diagram of a liquid helium cryopump.

from the cryopump is compressed and stored for subsequent reliquefaction.

At 20°K, a pressure of 10^{-11} torr is obtainable for all gases except neon, hydrogen, and helium. To avoid the expense of obtaining lower cryopanel temperatures, it is usually preferred practice to provide alternative means for removing these gases.

Fig. 36(*a*) shows the basic arrangement of one type of liquid-helium cryopump (Forth [39]). The chamber C is first pumped by conventional means to 10^{-3} torr or below. The cryopumping cold surface, cryocondenser or cryopanel consists of a spiral of copper tubing, S (Fig. 36*b*), of which one end is connected via a vacuum-jacketed feed-pipe to the liquid-helium reservoir, R. The other end of this spiral is joined via a second spiral forming a radiation shield to a regulating valve V and then to a mechanical rotary pump. The liquid helium is thereby sucked into spiral S, where it evaporates, cools the spiral, and then passes as gas through the radiation shield spiral to the mechanical pump into a container for gas recovery. Adjustment of the temperature of the cryopump spiral, S, is obtained by regulating the control valve, V. In its path from S to the mechanical pump, the waste gas passing through the radiation shield reduces it to an intermediate temperature of about 20°K. With large installations, however, it is more economical to cool this radiation shield by a separate circulation of liquid nitrogen.

A cryopump of this type with a pumping speed of 2,000 litre per sec would need to have a cryopumping surface area given approximately by equation (1.21) to be

$$2{,}000 = 62{\cdot}5A/\sqrt{M}$$

where p_2 is neglected compared with p_1. For nitrogen, $M=28$, so

$$A = 2{,}000/11{\cdot}7 = 17 \text{ cm}^2$$

A supply of about 0·5 litre of liquid helium per hour would be needed for such a cryopump, provided the initial pressure obtained by conventional pumping is 10^{-3} torr or preferably less.

Honig [40] has described an all-metal ultra-high vacuum system consisting of a sorption pump for backing a 25 litre per sec cold-cathode getter-ion pump, which evacuates the chamber to 10^{-8} torr. Liquid-helium cryopumping is then used to attain an ultimate pressure in the chamber of 10^{-10} torr, with a pumping speed for air of 800 litre per sec (Fig. 37).

The liquid helium is not continuously circulated in this design, but is in a container made from a 600 ml stainless steel beaker provided with a welded stainless steel top maintained within the pre-pumped vacuum chamber. This liquid-helium cryopump or trap is surrounded on all sides, except its base, by walls at liquid-nitrogen temperature to minimize radiation losses, and it is supported by a thin stainless steel tube of low thermal conductivity to reduce conduction losses. Further insurance against heat loss is provided by coating all interior surfaces of the trap assembly with a 0·0005 inch layer of silver. The stainless steel beaker itself is plated with a 0·015 inch layer of silver.

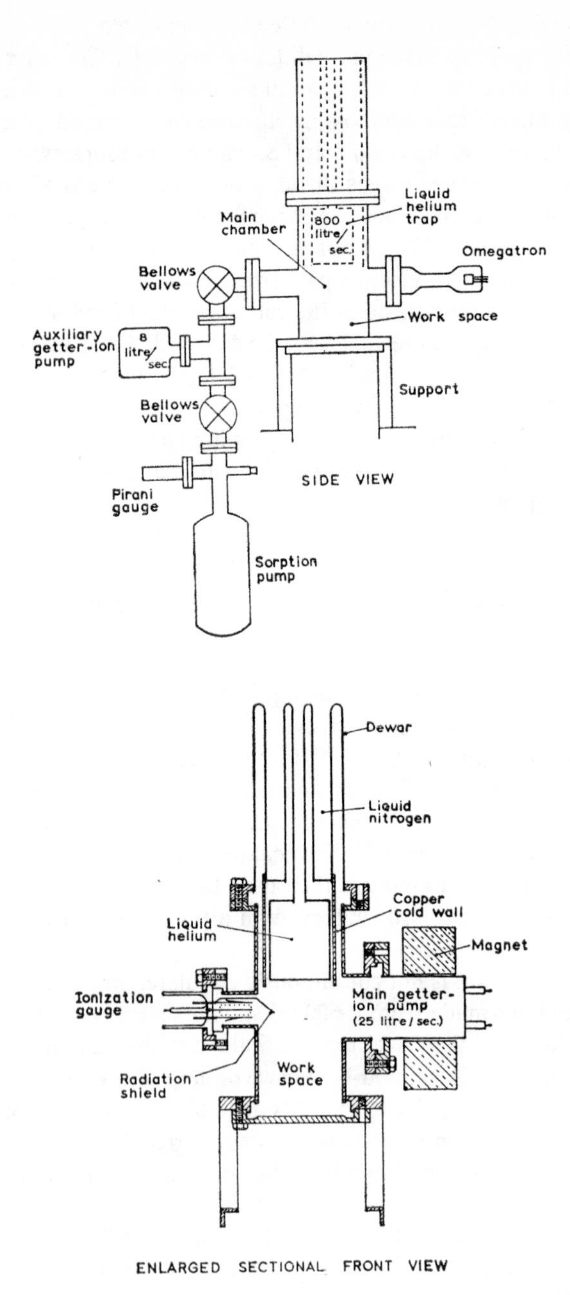

Fig. 37. Liquid helium cryopumped vacuum system due to Honig.

After preliminary pumping with the getter-ion pump and bake-out, the outer trap and the inner liquid-helium cryopump are first filled with liquid nitrogen. Preliminary cooling in this way conserves the expensive helium. When the inner helium cryopump or trap has reached −196°C, the liquid nitrogen is removed completely from it by blowing in helium gas. The liquid helium itself is then transferred to the inner trap. With the 600 ml beaker used, one full charge lasts about 4 hour.

The base of the liquid-helium-filled stainless steel beaker had a surface area of 58·5 sq cm. Its maximum theoretical pumping speeds were therefore (cf. equation (1.21), with $p_2 = 0$)

$$\text{for hydrogen:} \quad \frac{62{\cdot}5 \times 58{\cdot}5}{\sqrt{2}} = 2{,}600 \text{ litre sec}^{-1}$$

$$\text{for nitrogen:} \quad \frac{62{\cdot}5 \times 58{\cdot}5}{\sqrt{28}} = 696 \text{ litre sec}^{-1}$$

CHAPTER TWO

THE MEASUREMENT OF VACUA

2.1. *Introduction*

The scientist and engineer have always taken great pride in precision of measurement. The fundamental quantities, length, mass, and time, can all be determined to accuracies of better than 1 part in 10^8 if suitable apparatus is used, and 1 part in 10^5 is frequently met in the electrical sciences. The measurement of gas pressures with high precision, especially of those below 1 torr, is, however, not possible and it is doubtful whether it ever will be. The situation is increasingly difficult the lower the pressure. If a gas pressure below 10^{-4} torr is quoted as within $\pm 5\%$, the work has either been carried out with great care, or the error quoted is optimistic because various factors influencing accuracy have been overlooked.

Pressure is force per unit area and, as stated in section 1.1, the basic unit is the newton per sq metre (newton m^{-2}) or the dyne per sq cm (dyne cm^{-2}), though, generally, the mm of mercury or the torr are preferred in practice. However, a pressure of 10^{-6} torr corresponds to a difference in mercury levels of 10^{-6} mm, about 5 atomic diameters, so any direct observation in terms of hydrostatic head is out of the question. In the ultra-high vacuum range below 10^{-8} torr, the concept of a mercury level difference is meaningless except by inference! At a pressure of 10^{-3} torr, the force per sq cm is only 1·33 dyne, which can be measured with reasonable accuracy by balancing against a mechanical device or a mercury head, but, at 10^{-6} torr, the forces involved are very difficult indeed to measure mechanically.

Three further important sources of difficulty in pressure measurement are:

(*i*) that the gauge or its immediate surroundings may sorb or desorb gas, changing the conditions to be determined;

(*ii*) the gauge often cannot easily be placed in the system at the point where the pressure is required, and there may be an unknown pressure difference between this point and the gauge location;

(*iii*) the calibration of the gauge usually depends on the nature of the gas and the gas concerned may not be known.

Fortunately, in almost all the many processes undertaken *in vacuo*, it is not necessary to know the pressure accurately, only that it is below a certain value. For example, the residual gas pressure in an electron tube is low enough if the m.f.p. of the gas molecules is longer than a value which, depending on the density and cross-section of gas molecules available for ionization by electrons, makes it certain that ionization of the gas is small enough. If experience and theory indicate that a pressure p should not be exceeded to ensure satisfactory operation of the device or process, it is not necessary to place too much reliance on the gauge, but, if possible, ensure that a pressure below, say, $0{\cdot}1\ p$ is obtained. Even so, there are several instances where this procedure can be misguided and it is better to analyse the residual gases by a mass spectrometer. Thus, the pressure may be low enough if it is less than 10^{-5} torr, provided the bulk of the gas is not chemically-active, oxygen, say.

If the pressure and temperature are known, the molecular density n can be evaluated from $p=nkT$. This enables the rate at which molecules strike a surface in the vacuum to be calculated as $N=\frac{1}{4}n\bar{v}$, where $\bar{v}$ is $\sqrt{(8RT/\pi M)}$, provided the molecular weight, M, of the gas is known, i.e. the gas has been analysed. For work under ultra-high vacuum, the necessity for maintaining clean, gas-free surfaces is generally the prime requirement. Such surfaces will become covered with gas at a rate depending on $\frac{1}{4}n\bar{v}$. Knowing the partial pressure of the gases in the residual vacuum, this coverage can apparently be evaluated. But one also needs to know whether the molecules incident on the surface stick to it or not, and this can vary by a factor of at least 100 depending on the nature of the gas and the surface. In such a case, measurement of the pressure is not the only or even the chief requirement.

A final and considerable source of difficulty in pressure measurement is the problem of setting up a standard. Most vacuum gauges have a calibration depending on the nature of the gas and, further, this calibration, even for a given gas, cannot be related satisfactorily to the physical dimensions and characteristics of the gauge. It is essential, therefore, to calibrate such gauges against a standard. The U-tube oil and mercury manometers are available as standards at pressures above 1 torr. The McLeod gauge, in which the gas is compressed by a known ratio before its pressure is measured by a mercury manometer, is the *only* readily available absolute gauge of practical value in the range below 1 torr. But it exhibits increasing error at pressures below 10^{-3} torr and is hopelessly unwieldy for use below 10^{-6} torr.

2.2. *The Discharge Tube as a Vacuum Indicator*

A useful indication of the order of the pressure in a vacuum system can be obtained by setting up a glow discharge in the residual gas. A discharge tube (Fig. 38) is sealed onto a side-tube of the system. This sealing can be done by glass-blowing or by a hard-waxed conical joint or an O-ring (section 3.7).

The glass tube is preferably of Pyrex. D and C are two metal electrodes held inside this tube by means of glass-to-metal seals. The lead-in wires through these seals serve as terminals for the application of a high p.d. of 2 kV or more, preferably d.c., but a.c. from a high-voltage transformer may be used. The metal electrodes are made of nickel or stainless steel and are of generous thickness so that excessive temperature rise during the discharge is prevented.

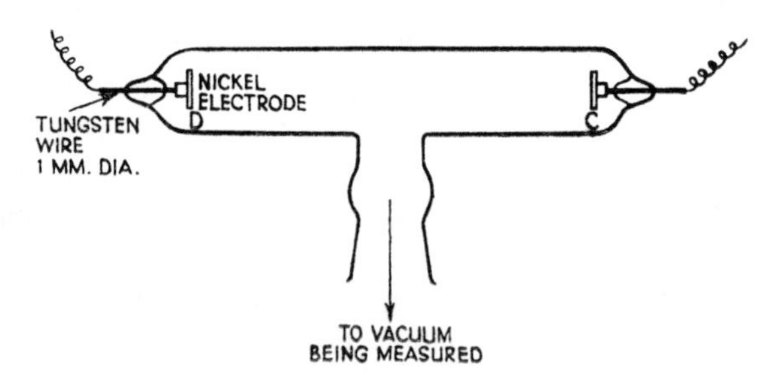

Fig. 38. The discharge tube.

A very convenient type of shielded discharge tube is supplied by Associated Electrical Industries Ltd.

The discharge obtained depends on the nature and the pressure of the gas, the current passing, and the electrode disposition and size. At a pressure of 1 to 20 torr, a streamer of discharge passes from one electrode to the other. At the higher pressures (order of 20 torr), this streamer is narrow. At about 1 torr, the streamer widens to the walls of the tube. As the pressure is still further decreased below 1 torr, definite regions in the glow discharge can be observed if a d.c. p.d. is used.

Surrounding the cathode there is the cathode glow. This takes up the contour of the cathode surface. There follows the Crookes' dark space and then the negative glow. From the anode extends the positive glow, which is either continuous or striated. The negative and positive glows are separated by the Faraday dark space.

The results give only the order of the pressure, serving as a rough indication. However, the appearances can be put on some sort of quantitative basis because the length of the Crookes' dark space, d, depends on the m.f.p., L, and therefore also on the pressure, p. The relationship is

$$p = \frac{k}{L} = \frac{k_1}{d} \tag{2.1}$$

k and k_1 being constants. This is a rough indication only, because the length d also depends to a small extent on the current density.

At pressures lower than 10^{-1} torr, the walls of the discharge tube begin to fluoresce. This fluorescence occurs, then, at what is usually a satisfactory backing pressure, i.e. one attained with oil-sealed mechanical pumps. The pressure at which this fluorescence begins also depends on the nature and age of the glass and on the applied voltage. At pressures less than 10^{-2} torr, the discharge disappears if the applied electric field is less than about 100 volt per mm: this state is called a 'black' discharge.

The discharge can also be excited by setting up in the gas a high frequency alternating electric field generated by an oscillator or by a Tesla coil. The latter is useful when a glass vessel is being pumped and the system has no discharge tube. The appearance of the glow indicates when a satisfactory backing pressure is attained, and that vapour pump heaters can be safely switched on. This state, corresponding to a pressure below 10^{-1} torr, is indicated by the glow pervading the whole system, without a trace of the pink, nitrogen discharge.

Table 2.1 shows how the colour of the discharge depends on the

TABLE 2.1

Variation of the colour of a discharge with the nature of the gas

Gas	*Cathode glow*	*Negative glow*	*Positive glow*
Air	Red	Blue to pink	Pink; blue at lower pressures
Ammonia	Blue	Green-yellow	Blue
Argon	Red-pink	Blue	Violet-red
Helium	Red	Pale-green	Violet-red to yellow-pink
Hydrogen	Brown-pink	Light-blue	Reddish-pink
Mercury vapour	Green	Yellow-white	Greenish
Neon	Yellow	Orange	Blood-red
Nitrogen	Red-pink	Blue	Reddish-yellow
Oxygen	Red	Green-yellow	Lemon-yellow with reddish core
Water vapour	White-blue	Blue	White-blue

gas in the vessel or discharge tube. This can be made a more scientific method by examining the discharge with a spectroscope. In the case of air, the colour of the discharge changes from predominantly pink to a pale blue or bluish-white as the pressure is decreased. This is because the relative proportions of the gases present in air change as the pressure is lowered. Any carbon dioxide content will increase, because this gas has a higher molecular weight and so a smaller flow rate to the pump than the nitrogen and oxygen. If an oil-sealed mechanical pump is used to evacuate the discharge tube, the condensable gases, chiefly water vapour, remain at the lower pressures. Finally, the ions in the gaseous discharge impinge on the tube walls and release gas from the glass. This gas is chiefly water vapour.

The discharge tube method is a reasonably satisfactory and easy means of indicating the pressure in unsophisticated vacuum plant, as used, for example, for the evaporation of metals *in vacuo*, the fractional distillation of chemicals, and the impregnation of electrical components, and in food-canning and vacuum drying techniques. If a discharge tube of known dimensions is used and with a certain d.c. or a.c. voltage across it, the point at which the discharge goes 'black' is frequently a satisfactory means of indicating an adequate vacuum. By using an adjustable spark-gap across the high-voltage supply as an indication of the p.d. applied, a more quantitative method is available. The higher the voltage at which a 'black' discharge is obtained, the lower the pressure.

With experience, impurity vapours in the gas can be identified to some extent: water vapour and grease vapour from stopcocks are made evident by characteristic whitish-blue and blue glows respectively. The presence of even a small percentage of impurity often completely spoils the correlation between the nature of the gas and the discharge indicated in Table 2.1.

2.3. *The McLeod Gauge*

The only straightforward, absolute, vacuum gauge for measuring gas pressures in the range down to 10^{-5} torr is that first introduced by McLeod [41] in 1874. Fig. 39 shows two convenient forms of the gauge, which is best made of a borosilicate glass such as Pyrex or Hysil; of the several grades of Pyrex available, the chemical ware 7740 is the most commonly used.

A glass bulb A is attached to a fine capillary tube B of bore diameter usually 1 mm approximately. C is a side-arm provided with a capillary at D, which has the same bore as the capillary tube B, and runs close to and parallel with it. The tube C extends upwards to a stopcock S or vacuum valve, which is then joined by tube E to the system in which the pressure, p, is to be determined. After construction by the glass-blower and before connecting to the system via the stopcock, the gauge, with the top of the capillary B open and without its mercury filling, is first thoroughly cleaned (p. 240). The volume of the bulb A is determined and also the diameter of the capillary tube

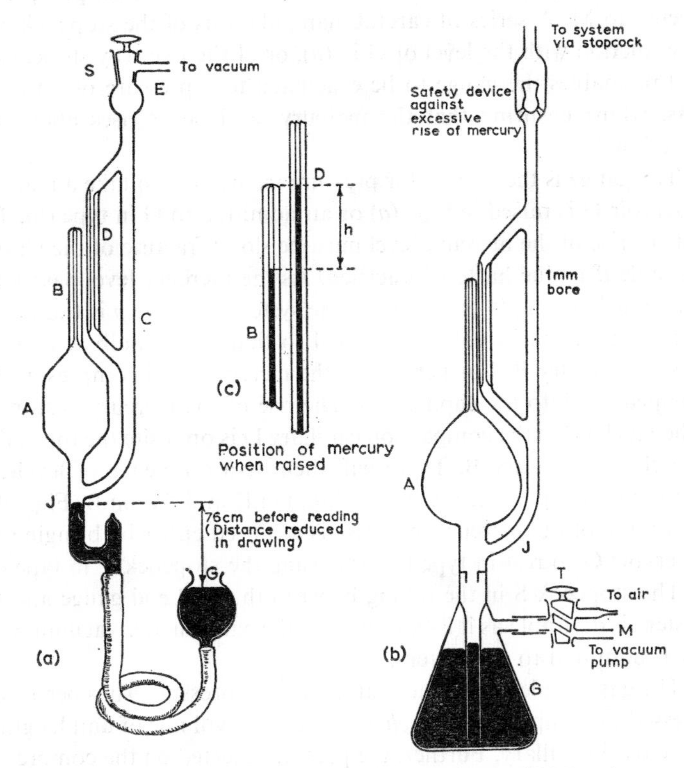

Fig. 39. (*a*) The McLeod gauge. (*b*) Shortened form of McLeod gauge.

B. The top of the capillary tube B is sealed-off by glass-blowing, to leave a flat inside top. The gauge is then mounted in a suitable support and joined by glass-blowing to the system with the stopcock S within the line from C to E. With the gauge filled with air at atmospheric pressure, clean, doubly-distilled mercury is then admitted to the reservoir G.

When first using the gauge, care has to be exercised to prevent

mercury from being splashed about during the initial evacuation. The usual practice is to close the stopcock S, establish a backing pressure in the system, open the stopcock S very carefully, meanwhile adjusting the height of the reservoir G to prevent the mercury level increasing above the junction at J. If the type (*b*) in Fig. 39 is used, the practice is to pump the gauge slightly by temporarily opening the stopcock S, then turn the two-way tap T to slightly evacuate the reservoir G by means of the small auxiliary mechanical pump connected to M. A series of careful manipulations of the stopcock S in conjunction with the level of G in (*a*), or of the two-way stopcock T in (*b*), enables the gauge to be evacuated to a pressure of 1 torr or less, whilst ensuring that the mercury level never rises above the junction J.

The gauge is then ready for pressure recording. To take a reading, reservoir G is raised in type (*a*) or air admitted to G in type (*b*). The rate of rise of the mercury level must be slow: fracture of the glass is possible if undue haste is practised. As the mercury level passes the junction J, a sample of the gas at the system pressure p is trapped in the gauge. It has a volume V equal to that of the bulb A plus the closed capillary B. Further rise of the mercury level compresses this trapped gas into the capillary B. The rise of the mercury is stopped when its level in the comparison capillary D is opposite the top end of the closed capillary B. There will then be a difference of levels, h, between the top of the mercury column in D and that in B (Fig. 39*c*). Cessation of the mercury level rise is achieved either by bringing the reservoir G to rest in type (*a*) or closing the stopcock T in type (*b*).

The stopcock S in the tubing between the McLeod gauge and the system is desirable as it is best to leave the gauge under vacuum when air is admitted to the system.

The gas sample of volume V at the system pressure p has been compressed to occupy a volume vh, where v is the volume of unit length of the closed capillary. Further, the pressure exerted on the compressed gas is clearly p plus that due to a head of mercury of height h.

Applying Boyle's law

$$pV = (p+h)vh$$

where p is in mm of mercury (torr) and h is also in mm. As p is usually much smaller than h, it can be neglected in $(p+h)$, so that

$$p = \frac{vh^2}{V} \tag{2.2}$$

where p is given in mm Hg (torr), v and V being in the same units (e.g. cu cm) and h in mm.

The pressure is therefore indicated on a square-law scale, which can be constructed simply from a knowledge of V and v, where the latter is $\pi d^2/4$, d being the capillary diameter.

Typical, convenient values are $V=300$ cu cm and $d=1$ mm, so that

$$\frac{\pi(0{\cdot}1)^2}{4} = 0{\cdot}00785 \text{ cm}^3$$

Therefore, $v/V=2{\cdot}6\times10^{-5}$. The range of pressures recordable, if the total length of the closed capillary is just over 100 mm, is therefore from $p=2{\cdot}6\times10^{-5}\times100^2=2{\cdot}6\times10^{-1}$ torr, for $h=100$ mm, to $p=2{\cdot}6\times10^{-5}\times1^2=2{\cdot}6\times10^{-5}$, for $h=1$ mm.

It is assumed here that a value of h less than 1 mm cannot be recorded accurately. With a high vacuum of 10^{-6} torr or less, the mercury will virtually fill the capillary completely. On allowing the mercury to drop to take a fresh reading, surface tension will cause it to stick against the closed end of the capillary B, until the weight of the falling mercury pulls it away; this is referred to as a 'sticking' vacuum. Often, it is a good indication of a satisfactorily low pressure, which is less than 10^{-5} torr but has not been measured!

By choosing various values of the ratio v/V, different pressure ranges can be obtained. It is not possible to make d less than 0·5 mm, and even then trouble will be experienced with fracture of the mercury column in the closed capillary on lowering the mercury. Further, V can only be increased above 500 cu cm if a support (e.g. in the form of a carefully fitted plaster of Paris mould) is maintained around the bulb A to sustain the weight of the mercury. V has been made 1,300 cu cm in some designs, but they are unwieldy and the mercury is expensive.

The McLeod gauge has been much used in vacuum technique, and the general practice has been to calibrate all other gauges against it. Its accuracy below 10^{-4} torr is poor, however, so an alternative in the form of a standard orifice is gaining favour as the primary standard for gauge calibration in the vacuum range below 10^{-4} torr (section 3.9).

The outstanding advantages of the McLeod gauge are cheapness (apart from the cost of the mercury), simplicity of operation, and, particularly, the fact that it can be easily calibrated in terms of its readily-measured physical dimensions, and that this calibration is independent of the nature of the gas, provided it is not a vapour which condenses to a liquid during the compression.

There are unfortunately several **disadvantages of the McLeod gauge,** especially for measurements below 10^{-4} torr. In the summary of these given below, practice for minimizing errors where possible is discussed.

(*a*) It will not register correctly the pressure due to condensable gases. Water vapour is often present in air. Apart from the fact that such a vapour does not obey Boyle's law (assumed in arriving at the calibration), if the water vapour content of the air is high, it will condense to liquid during the compression. For this reason, the ultimate pressure attained by a mechanical pump, when recorded by a McLeod gauge, will appear significantly lower than that recorded by, for example, a Pirani gauge (section 2.4), which responds to the total pressure due to both permanent gases and condensable vapours. A means of removing the water vapour (e.g. gas-ballast or cold trap) may be used, but the readings of the two gauges will still be different, because of the vapour pressure of the rotary pump oil, particularly if this oil is contaminated.

(*b*) The McLeod gauge cannot record pressure instantaneously, and therefore is unsatisfactory for recording pressure changes except over long periods. The reasons for this are, first, time is needed to raise the mercury in taking a reading and, second, the gauge is inevitably connected to the system by a fair length of narrow tubing, so that it usually takes about a min for the gauge and system pressures to equalize because of the restricting effect of tubing on the flow of gases (section 3.3).

(*c*) The presence of the mercury in the gauge introduces a vapour pressure of about 10^{-3} torr at room temperature. The total pressure in the system cannot therefore be reduced below this value unless a cold trap at liquid-nitrogen temperature or the temperature of solid carbon dioxide is introduced between the gauge and the system.

A suitable arrangement of such a cold trap is shown in Fig. 40. This cold trap in itself introduces errors in pressure reading which are particularly important at pressures less than 10^{-4} torr. This intrinsic error is due to a pressure difference Δp in the gas in the tubing connecting the gauge to the cold trap. This pressure difference results from the continuous flow of mercury vapour from the gauge to the cold trap, which drives some of the gas molecules before it, so making the pressure in the gauge less than the pressure near the trap. The effect is similar, though of much smaller magnitude, to that in the action of the mercury diffusion pump, and was first pointed out by Gaede [42]. It was not for many years later, in 1961, however, that Ishii and Nakayama [43] gave clear experimental evidence that an important source of error resulting from mercury diffusion had hitherto been overlooked. Their findings were later substantiated by Meinke and Reich [44] in 1962. Sometimes known as the 'Ishii effect', the pressure difference Δp can be calculated to a sufficient degree of accuracy, in the case of a long cylindrical tube of radius R cm at

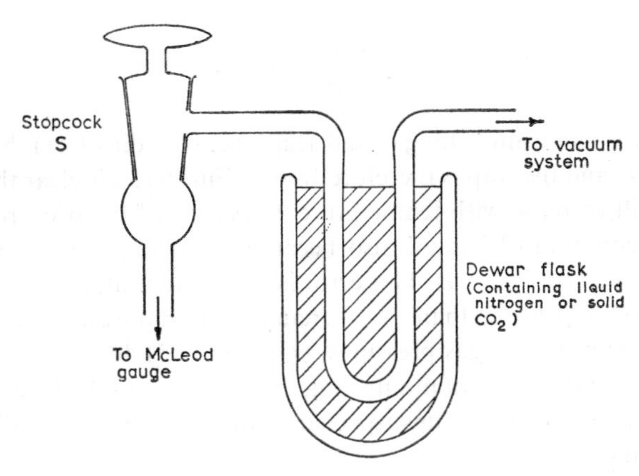

Fig. 40. Use of a cold trap between a McLeod gauge and a system.

molecular pressures, from an equation given by Ishii, following Gaede,

$$\log_e\left(1+\frac{\Delta p}{p}\right) = 0{\cdot}9Rp_m\sqrt{(T)}/D \qquad (2.3)$$

where p is the mean gas pressure in the tube, p_m is the vapour pressure, in torr, of mercury at the temperature T°K in the gauge (i.e. room temperature), and D is the diffusion coefficient in mercury vapour of the gas concerned, at T°K and atmospheric pressure; p is less than p_m.

At 20°C, i.e. 293°K, p_m is $1{\cdot}2\times10^{-3}$ torr and the diffusion coefficient (which decreases by about 10% per 10°C drop in temperature) for nitrogen in mercury vapour is 0·108 sq cm per sec. This is the value at 760 torr; the value required at p torr is obtained by assuming that the diffusion coefficient is inversely proportional to the pressure, which is included in the numerical factor given in equation (2.3). Substitution of these values in equation (2.3) gives

$$\log_e\left(1+\frac{\Delta p}{p}\right) = 0{\cdot}9R\times 1{\cdot}2\times 10^{-3}\sqrt{(293)}/0{\cdot}108 = 0{\cdot}17R$$

For $R = 1$ cm,

$$2{\cdot}3\log_{10}\left(1+\frac{\Delta p}{p}\right) = 0{\cdot}17$$

$$\therefore\ 1+\frac{\Delta p}{p} = 1{\cdot}186$$

i.e.

$$\frac{\Delta p}{p} = 0{\cdot}186$$

The error in using a long cylindrical tube, of radius 1 cm, between the gauge and the trap is therefore 18·6%. Further, it is clear that this error will increase with temperature T because of the occurrence of $\sqrt{T}$ in equation (2.3), and also because p_m increases more rapidly with temperature than D. This error is serious for values of the mean gas pressure p lower than the mercury vapour pressure, i.e. below 10^{-3} torr; at higher gas pressures, equation (2.3) does not apply.

There are two ways of minimizing this error due to the Ishii effect – which has led to many unsuspected erroneous pressure readings in past years – these are:

(*i*) to cool the McLeod gauge with solid carbon dioxide at the junction J in Fig. 39, so as to reduce p_m, the vapour pressure of mercury (this method is discussed by Ishii and Nakayama);

(*ii*) a more practical way is to introduce a short section of small bore tubing within the main connection between the gauge and the trap.

If this short section is, say, 1 cm long and 0·1 cm in radius, at 20°C, it is seen from equation (2.3) that the percentage error becomes only 1·8% approximately. The impedance to gas flow of this section of narrow tubing will, however, increase the time taken for the gauge and system pressures to become equalized.

(*d*) A significant error can occur in the reading of low pressures because the capillary depression of the mercury in the capillary tubes causes uncertainty in the measurement of h (equation 2.2). This arises because of considerable variation in the angle of contact between the mercury and the glass wall of the capillary tube, depending on whether the mercury is raised or lowered to its final position in taking a reading. Unfortunately, this error increases as the capillary bore diameter is reduced below 1 mm.

A method for reducing the variation in contact angle, and therefore the error, is to grind the inside wall of the capillaries, to roughen them. Even so, according to Leck [45], following careful measurements made by Rosenberg [46] who used a McLeod gauge with a bulb diameter of 1,300 cu cm and a very uniform roughened capillary of diameter 0·63 mm, the error amounts to $\pm 6\%$ at 10^{-5} torr. With a small bulb volume, the error will increase in proportion. Tapping helps to reduce the error due to capillary depression but does not eliminate it.

Non-uniformity of the capillary bore diameter also introduces an obvious source of error. Jansen and Venema [47] recommend selected capillary tubing of diameter uniform to within 1%.

(*e*) It is virtually impossible to provide a square top to the closed capillary by direct glass-blowing. This introduces an error in the measurement of small values of h, important at pressures below 10^{-4} torr. Further, if the capillary inner walls are roughened (e.g. by carefully drawing through the capillary a wire loaded with fine carborundum or diamond powder), direct sealing-off the top end will flame-polish the upper capillary wall. To avoid this difficulty, Podgurski and Davis [48], following Barr and Anhorn [49], inserted an 8 mm length of tapered glass plug into the top of the initially open capillary and then sealed with a flame around the capillary tip.

The *Vacustat* is a commercial form of swivel-type McLeod gauge marketed by Edwards High Vacuum Ltd. It follows a design originated by Flosdorf [50]. It is very convenient for measuring gas pressures in the range from 10^{-2} to 10 torr, requires only about 8 cu cm of mercury for filling, and avoids the use of a movable reservoir of mercury or an auxiliary vacuum pump. As depicted in Fig. 41, the overall height including a tripod stand is only 9 inch. The system in which the gas pressure is to be measured is connected by flexible rubber pressure tubing to the union at A behind the mounting panel; alternatively, a greased cone-joint in glass tubing or a rotary O-ring seal to metal tubing may be used. Bodily rotation of the Vacustat through a right angle about A is then possible. Before measuring the pressure, the gauge is in the horizontal position (Fig. 41*a*) so that the gas fills the gauge at the correct pressure. On requiring to take a pressure reading, the gauge is simply rotated to the vertical position (Fig.

41*b*), when the mercury runs out of the container B, compressing a sample of the gas contained within the region CD to a short length CE of the closed capillary tube. By arranging that the level in the comparison capillary F is opposite the end of the graduated capillary CE, it is possible to employ a square-law scale behind this capillary, which registers the pressure directly. A slight adjustment of the gauge about the vertical position is needed to adjust the mercury level correctly. The method of calibration is the same as for the McLeod gauge.

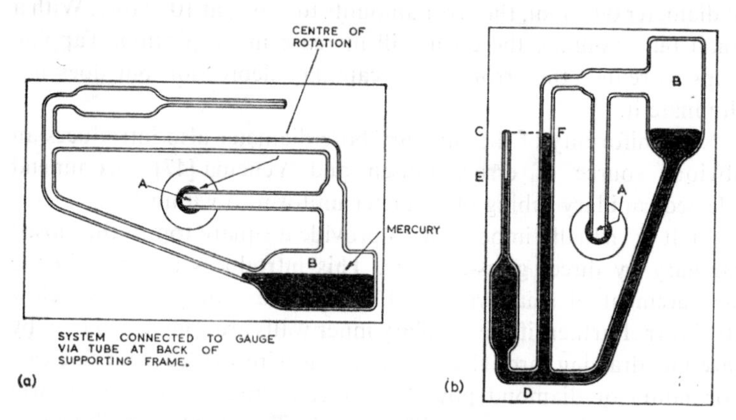

Fig. 41. The Vacustat (Edwards High Vacuum Ltd.).

2.4. *Thermal Conductivity Gauges*

For a given gas at sufficiently low pressures, the thermal conductivity decreases with pressure. This principle is utilized in vacuum gauges of which the essential arrangement consists of a thin metallic wire or filament of diameter d, maintained along the axis of a tube of much larger diameter D, the open end of which is connected to the system in which the gas pressure p is to be measured. The axial wire is heated to about 200°C by the passage of electric current. The rate at which heat from this wire is conveyed through the gas to the tube wall at ambient temperature decreases as the gas pressure decreases. The temperature of the wire is therefore dependent on this gas pressure, rising as the pressure falls. In the Pirani [51] gauge, the change of resistance of the wire with temperature and consequently with pressure is measured by a Wheatstone bridge. In the first place, therefore, the gas pressure is measured in terms of the wire resistance. In the thermocouple gauge, the temperature of the wire is recorded directly by an attached thermocouple: a calibration of thermocouple e.m.f. against pressure is obtained.

A simplified theory of operation is valuable in determining relevant operating characteristics. Suppose the central wire is heated by the passage of electrical current to a temperature T°K, whilst the inner wall of the surrounding tube is at the ambient temperature T_A°K; where $T > T_A$. The wire is very thin: 0·05 to 0·1 mm in diameter, whereas the tube has a diameter of 1 cm or more. Gas molecules in the gauge tube will therefore make many more collisions with the tube wall than with the wire. It is a reasonable approximate assumption, therefore, that the gas molecules are at the temperature T_A and only attain the higher temperature T when they strike the wire. It is further assumed that on striking the wire, the molecules attain an energy corresponding to the temperature T.

The rate of transfer of thermal energy E from the heated wire to the gas is given by

$$E = kN(T-T_A)$$

where N is the number of molecules impinging on the wire per sec and k is a constant for a particular gas and wire surface. In the case where the gas molecules impinging on the wire arrive from distances not exceeding their m.f.p. at the pressure p of the gas, the number of molecules incident on the wire surface per sec is given through equation (1.9) as $pSN/\sqrt{(2\pi MRT_A)}$, where S is the surface area of the wire. Hence,

$$E = \frac{KpS}{\sqrt{(MT_A)}}(T-T_A) \tag{2.4}$$

where K is a second constant depending on the nature of the gas and the wire surface, and it assumed that all molecules have the wall temperature T_A before they strike the wire.

If the m.f.p. of the gas molecules, L, does not exceed the wire diameter d, many molecules which leave the wire after striking it will return to it again instead of passing into the gas towards the wall; the equation (2.4) is then invalid. Indeed, at the higher pressures, the energy transfer becomes independent of the pressure unless T is increased significantly. In general, the heat transfer per sec through the gas approximately follows equation (2.4) for gas pressures when L is larger than d. For a wire diameter d of 0·1 mm, the pressure range for nitrogen is therefore given by equation (1.4) as below 5/0·01 millitorr, i.e. below 0·5 torr.

The heat transfer at a given pressure p and for given values of T and

T_A varies inversely with the square root of the molecular weight M of the gas. It will also depend on K, which is decided by the nature of the gas (in particular the ratio of its specific heats at constant pressure and constant volume) and the wire.

In addition to the heat transfer away from the hot wire due to molecular motion in the gas, there will be two other sources of thermal loss: radiation of heat from the wire and conduction through the metal leads to the ends of the wire. Both of these are independent of pressure. The radiation loss becomes significant compared with the heat energy transfer through the gas at a pressure of about 10^{-5} torr. The conduction loss through the leads will be less the longer the wire

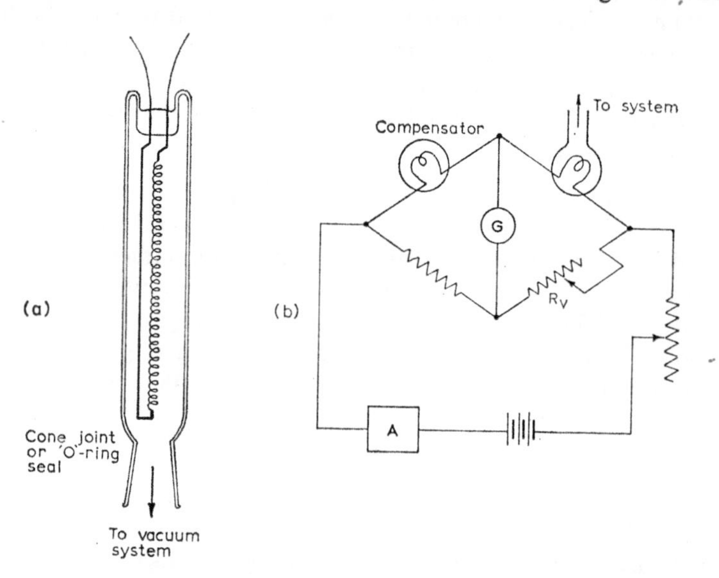

Fig. 42. (*a*) Pirani gauge head. (*b*) Basic Pirani gauge circuit.

filament used. But an overlong filament will be difficult to support without mechanical vibration, which will introduce wire-to-gas energy changes. These factors decide that the minimum pressure recordable by a thermal conductivity gauge is about 10^{-4} torr.

A typical **Pirani gauge** head and basic measuring circuit are shown in Figs. 42(*a*) and (*b*) respectively. The head consists of a glass tube of diameter 1 cm or more, with a greased cone-joint or more usually an O-ring connection (section 3.7) at the open end to the vacuum system in which the pressure is to be recorded. The Pirani filament, i.e. axial wire, is of tungsten, nickel, or platinum wire, 0·005 to 0·1 mm in diameter. To accommodate a length of 20 to 30 cm, it is usually wound into a helix of axial length, about 8 cm, and outside diameter 0·5 to 2 mm, with a pitch of at least 10 wire diameters to prevent any one turn from shielding its neighbours. The temperature of the wire is maintained at an operating temperature of about 200°C at the lowest pressure.

In the Wheatstone bridge circuit, a compensating head is often used, which is identical with the measuring head but sealed off under vacuum. This is placed in the arm of the bridge adjacent to that containing the measuring head, with the two heads in proximity, so that they are subjected to the same ambient temperature changes. A resistance may be used in the bridge instead of the compensating head but variations of room temperature are then more troublesome.

Instead of measuring the resistance of the Pirani filament, the usual procedure is to maintain constant the current to the bridge or the voltage across it from the supply, then balance the bridge by the variable resistance R_V when the pressure within the measuring head is at 10^{-4} torr or below. This gives the lowest pressure reading, marked zero, on the galvanometer G. As the pressure in the measuring head is increased its filament resistance decreases, the bridge is put out-of-balance and the galvanometer indicates the off-balance current. The galvanometer reading is then a measure of the pressure; a calibration graph of this deflection against pressure is obtained by measuring the pressure in the system (to which the Pirani gauge is attached) by means of a McLeod gauge.

A typical, modern, commercial Pirani gauge has a head with four filaments (Edwards High Vacuum Ltd.), with the necessary balancing coils for the Wheatstone bridge wound on a former placed around the gauge head (Fig. 43). This assembly is housed in a robust plastic or metal cylindrical container. Two of the filaments are in a pair of tubes which connect to the vacuum system via a greased cone-joint or an O-ring. The other two comparison filaments are sealed-off in separate tubes. All the filaments are of the same construction. The two measurement filaments are in opposing arms of a Wheatstone bridge with the two comparison filaments in the remaining arms. A multi-wire cable joins the gauge head to the mains or battery-operated unit which supplies the necessary electrical power and incorporates a meter (which registers the out-of-balance bridge current) directly calibrated in pressure units for dry air. By changing gauge heads, the pressure ranges can be: (*a*) 10 to 5×10^{-3} torr, using an atmospherically balanced bridge, i.e. the two compensating filaments are sealed-off in tubes at atmospheric pressure; or (*b*) 1 to 10^{-4} torr, with a vacuum balanced bridge in which the compensating filaments are sealed-off in tubes at low pressure.

To ensure that the voltage across the bridge circuit is constant irrespective of supply mains or battery voltage fluctuations, a transistorized constant voltage power-pack with a very low output impedance of 0·1 ohm, providing a p.d. across the bridge of 10 volt

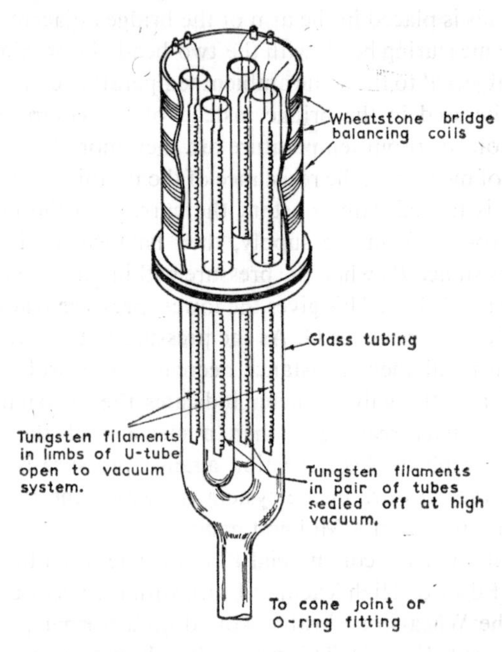

Fig. 43. A commercial Pirani gauge head with incorporated bridge balancing coils.

approximately, is used. This p.d. will then remain constant irrespective of changes in the load due to variations of resistance of the two Pirani filaments with pressure.

Using a simpler circuit with a 10 volt accumulator supply, in which the current through a single Pirani filament was kept constant and not the voltage across the bridge, the calibration graph shown in Fig. 44 was obtained.

Reconditioning of Pirani filaments is sometimes necessary. These gauges are frequently used in the pressure range from 10^{-4} to 1 torr on plant where the vacuum conditions are 'dirty', the system being exposed to water vapour and oil vapour. If the Pirani filament surface becomes excessively contaminated, there will be variation in the manner in which the energy of impinging gas molecules becomes accommodated to the temperature of the filament. This can cause a significant error in pressure reading. To clean the filament, current is passed through it for about 1 min to raise its temperature to some 900°C. A switch on the control unit is arranged to connect the filament to the supply so that about four times the normal operating current is passed through it. The operating temperature of about

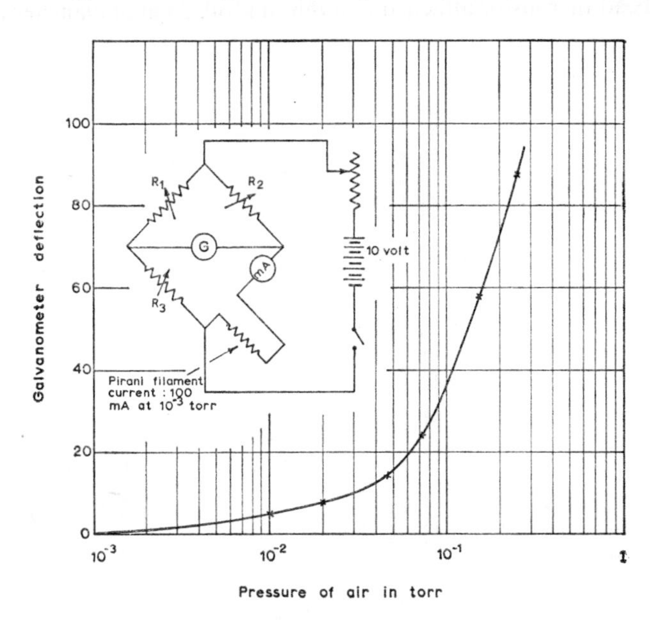

Fig. 44. A plot of galvanometer deflection against air pressure for a Pirani gauge head with a tungsten filament of total length 30 cm and wire diameter 0·03 mm.

200°C is then restored before using the gauge for pressure measurement.

An interesting alternative to the use of Pirani filaments of metallic wire is the use of thermistors, in the **thermistor gauge**. Thermistors are mixtures of oxides, including those of iron, nickel, copper, zinc, and manganese, which are subjected during manufacture to a specialized heating and sintering treatment. They have large negative temperature coefficients of resistance of between 1 and 6% per degC and are commercially available in the form of small beads, rods, or plates.

The thermistor was first used for pressure measurement by Meyer [52], and Weise [53]; later by Becker, Green and Pearson [54], and also Gruber [55]. One of the most recent exponents of this practice is Varićak [56]. Following his reasoning, it is pointed out that if S, the surface area of the wire or thermistor, in equation (2.4), is increased, greater energy transfer takes place through the gas for a given temperature difference $(T-T_A)$. In practice, however, plate thermistors introduce too large a time lag between change of pressure and change of gauge response. Varićak therefore recommends the use of a miniature bead thermistor attached to a thin tin foil, 3 cm in diameter and

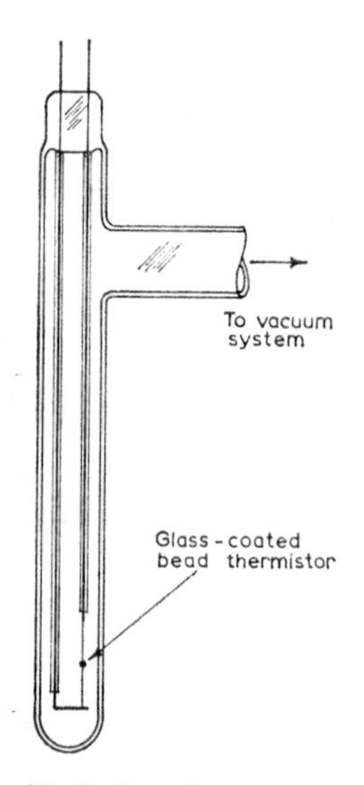

Fig. 45. A thermistor gauge head.

0·01 mm thick, able to record pressures of air over the range from 10^{-1} to 5×10^{-6} torr.

Roberts, McElligott and Jernakoff [57] have developed a thermistor gauge, for the pressure range 10^{-4} torr to 1 torr, based on a glass-covered bead thermistor (General Electric type 1B153), which is spot-welded to glass-covered Fernico or tungsten leads in a 1 cm diameter glass tube (Fig. 45). This tube can be baked at 450°C (some types of thermistor will not withstand heating above 200 to 300°C). The thermistor bead is heated to 100°C by the passage of current when its resistance is rather less than 1,000 ohm. The tube can be thermostated, if required, in an ice-water bath at 0°C. The thermistor is incorporated in a Wheatstone bridge circuit with 800 ohm resistance arms. The full circuit arrangement included a transistorized, stable, voltage supply for this bridge, and balance recording by a differential amplifier. The glass-covered thermistor at 100°C is reported to have particularly good inertness to gaseous atmospheres.

The **thermocouple gauge** operates, like the Pirani and thermistor gauges, on a basis of the change with pressure of the thermal transfer through a gas between two surfaces at different temperatures, but

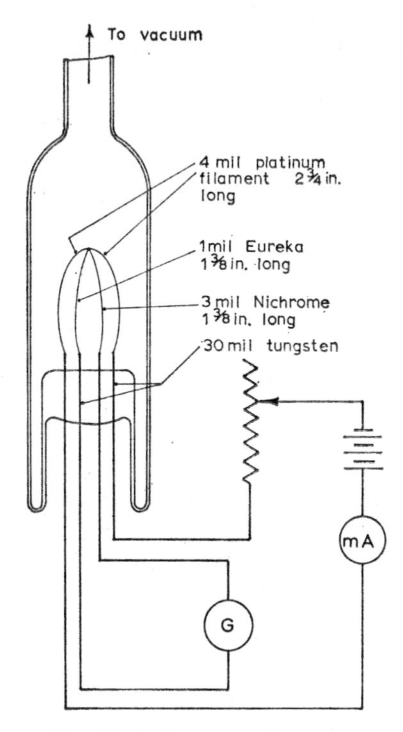

Fig. 46. A thermocouple gauge.

employs a heated filament of wire to which a thermocouple is attached to monitor its change of temperature with pressure.

First introduced by Voege [58], a modern form of the thermocouple gauge is shown in Fig. 46. An alternative, convenient pattern is based on a Best Electrics type S50 vacuum thermocouple (actually designed for radio-frequency current measurements), which has a heater resistance of 125 ohm and takes a current of up to 10 mA. The attached thermocouple resistance is 8 ohm and gives an open circuit e.m.f. of 20 mV at high vacuum, with a heater current of 10 mA. The operating circuit using a battery supply is very simple: a p-n-p junction transistor, with its particularly-constant collector current irrespective of collector-base voltage variations, can be used to ensure a constant current of, say, 8 mA through the heater wire, where this current is adjusted by the emitter current control provided by the variable resistance (Fig. 47*a*). A pointer moving-coil galvanometer (0 to 300 μA; resistance 100 ohm) was used to record the thermocouple e.m.f. A calibration curve of pressure against galvanometer deflection for air over the pressure range 0·001 to 1·0 torr is shown in Fig. 47(*b*).

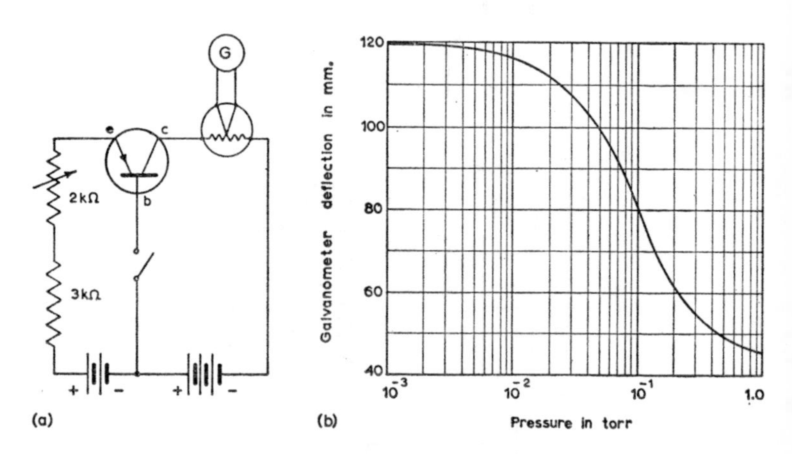

Fig. 47. (*a*) Operating circuit for a thermocouple gauge. (*b*) Calibration graph of pressure against galvanometer deflection (for air).

A commercial model of thermocouple gauge (Genevac Ltd. type TCG2) covers the pressure range from 5 to 10^{-3} torr and employs a.c. heating. Heating by a.c. offers an advantage over d.c., as the latter may cause stray e.m.f.'s to be developed across resistance at the junction between the thermocouple and the heater.

The advantages of thermal conductivity gauges are:

(*i*) small volume, easily attached to the system with electrical operation from a control unit which can, if required, be remote from the gauge head and also serve, by selector switching, several gauge heads;

(*ii*) the time lag between change of pressure and change of gauge response is negligible for most practical purposes;

(*iii*) the output e.m.f. can be readily amplified (transistor amplifiers are preferable because of the comparatively low output impedance of the gauge), enabling display on a recorder to be achieved so that a continuous record of pressure variation with time is possible;

(*iv*) as compared with the McLeod gauge, thermal conductivity gauges indicate the total pressure in the system due to both permanent and condensable vapours.

Thermal conductivity gauges are frequently employed to record the backing pressure in a vacuum system and to monitor the pressure in a plant in which the total pressure is between 10^{-3} and 5 torr. Their electrical output can be used to operate a relay to an electrically-operated vacuum valve, enabling semi-automatic plant to be constructed in which valve operation at a predetermined pressure is obtained.

The disadvantages of thermal conductivity gauges are:

(*i*) the lowest recordable pressure is about 10^{-4} torr. This can be extended by a decade if the gauge is carefully operated with a constant current to the filament and thermostated walls. In general, commercial gauges are not recommended for use below 10^{-3} torr;

(*ii*) the calibration depends on the nature of the gas. Roughly, the thermal transfer through the gas varies inversely as $\sqrt{M}$, M being the molecular weight of the gas, as shown by equation (2.4), so the gauge response will be larger for hydrogen than for nitrogen. This is, in fact, an advantage in that it enables the gauge to be used for leak-finding (section 3.11), but the chief difficulty is that the composition of the gases in a vacuum system is often not known, so the pressure recording (assuming calibration for dry air) will be considerably in error;

(*iii*) the gauge calibration will alter if the filament becomes contaminated. In this respect, the thermocouple gauge is probably a better choice than the Pirani gauge and, as reported above, the small glass-covered thermistor type can be designed to minimize this error.

2.5. *The Knudsen Gauge*

The operation of the Knudsen [59] gauge depends on the variation with pressure of the momentum of impact of heated gas molecules on a light, suspended vane. The heating is usually arranged by an electrically-heated filament at a temperature T_1 near the vane, which is suspended from a torsion fibre and is at a lower temperature T in the gas. The vane is deflected by the couple exerted by the impinging gas molecules. This deflection is a measure of the pressure for given values of T_1 and T and is largely independent of the nature of the gas. In this sense, the gauge is an absolute manometer. It is not readily possible, in practice, to determine T_1 and T with precision, and it is difficult to make the gauge absolute in the wider sense that the pressure is known solely from a knowledge of the physical parameters involved. Indeed, many attempts have been made to develop this theoretically-attractive gauge for general use, but they cannot be said to have succeeded, as there is no commercial model generally available, and the gauge has been used mostly in specialized research.

The theory of working may be explored by considering a fixed flat plate A, heated to a temperature T_1°K (i.e. electrically heated by a filament), near a parallel flat vane B which is pivoted about an axis O. B is at the lower temperature T°K, which is assumed for simplicity to be the same as that of the surrounding cylindrical container C, which is at room temperature (Fig. 48). The separation between the parallel

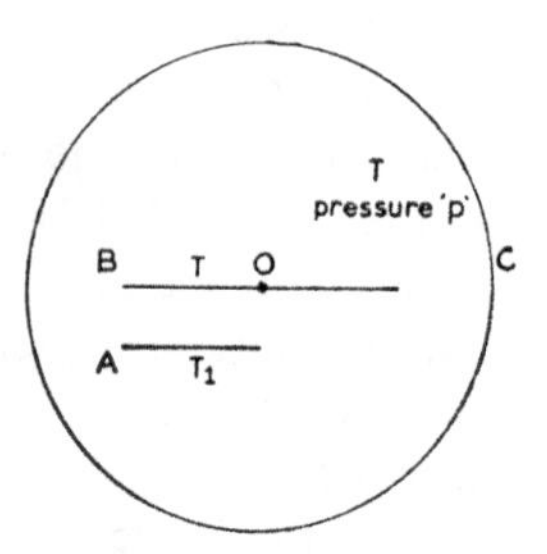

Fig. 48. The principle of the Knudsen gauge.

plates A and B is assumed to be small compared with their linear dimensions, so that edge effects can be neglected, and also small compared with L, the m.f.p. of the gas molecules at the prevailing pressure. Molecules which leave plate A will consequently arrive at plate B without any change in their condition.

Let m be the mass of a molecule of a given gas in the gauge. The molecules which leave A at the temperature T_1 will have a Maxwellian distribution of velocities. Suppose there is a group of them comprising n_1 molecules per unit volume with velocity components u_1, v_1, and w_1, in directions along three mutually perpendicular axes x, y, and z, respectively; where x is the direction perpendicular to B. The number of such molecules which strike unit area of B per sec will be half the number in a cylinder of unit cross-section and length u_1 perpendicular to B; this will be $n_1u_1/2$. (The division by 2 is necessary because only half the molecules will leave A in the positive direction towards B; the other half will leave the back face of A to enter the space in the gauge outside the region confined between A and B.) Each of these molecules will impart a change of momentum of $2mu_1$ to B. The change of momentum per unit area at B per sec due to this group will therefore be $n_1mu_1^2$. The total change of momentum per sec due to all the molecules from A which reach B will therefore provide a force F_B on B, given by

$$F_B = am \sum n_1 u_1^2 = amn_A\bar{u}^2 \tag{2.5}$$

where a is the area of vane B, n_A is the total number of molecules per unit volume which leave A to travel towards B, and $\bar{u}^2$ is the mean square of the velocity components along the x-axis given by

$$\bar{u}^2 = \frac{n_1u_1^2+n_2u_2^2+n_3u_3^2+\cdots}{n_1+n_2+n_3+\cdots}$$

where n_2 molecules per unit volume have x-velocity components, u_2; n_3 have u_3, etc.

If C_A is the *root mean square velocity* of the gas molecules leaving A, defined by

$$C_A = \left(\frac{n_1c_1^2+n_2c_2^2+n_3c_3^2+\cdots}{n_1+n_2+n_3+\cdots}\right)^{1/2}$$

where c_1 is the resultant of u_1, v_1, and w_1, whilst c_2 is the resultant of u_2, v_2, w_2, etc., and it is assumed that $\bar{u}^2=\bar{v}^2=\bar{w}^2$, then

$$\bar{u}^2 = \tfrac{1}{3}C_A^2$$

So equation (2.5) becomes

$$F_B = \tfrac{1}{3}amn_AC_A^2 \tag{2.6}$$

Correspondingly, there will be a force F_A exerted on A by molecules leaving B at the temperature T, given by

$$F_A = \tfrac{1}{3}amn_BC_B^2 \tag{2.7}$$

where C_B is the root mean square velocity of the molecules at temperature T, and n_B is the number of these molecules per unit volume which leave B and travel towards A.

There will also be a force exerted on the back face of vane B facing away from A, due to the pressure p in the gauge. Hence, the force F on B is decided by

$$F = a(\tfrac{1}{3}mn_AC_A^2+\tfrac{1}{3}mn_BC_B^2-p) \tag{2.8}$$

because F_A and F_B from equations (2.6) and (2.7) will be additive, as plate A is fixed.

Two further relationships between C_A and C_B are obtainable. In the first place, molecules with the mean velocity $\bar{v}_A$ will reach the back face of B from the gas *outside* the interspace AB at a rate decided by $\frac{1}{4}n_A\bar{v}_A$ (equation 1.8), whereas molecules with mean velocity $\bar{v}_B$ will leave this face at a rate $\frac{1}{4}n_B\bar{v}_B$. These two numbers must be equal, otherwise there would be an accumulation of gas on this back face of B. Also, as the mean velocity is related to the root mean square velocity by a numerical constant,

$$n_A C_A = n_B C_B \tag{2.9}$$

In the second place, consider unit area of a boundary in the gas separating the interspace between A and B from the gas at pressure p in the container. Let n be the number of molecules per unit volume in this outside space at the pressure p. These will cross this boundary into the interspace at a rate $\frac{1}{4}n\bar{v}_B$, assuming the gas is at the same temperature as B; moreover, molecules from the interspace will cross this boundary towards the outside space at the rate $\frac{1}{4}(n_A\bar{v}_A+n_B\bar{v}_B)$. Therefore,

$$n\bar{v}_B = n_A\bar{v}_A+n_B\bar{v}_B$$

because there is an equilibrium in the gauge without any accumulation of gas between A and B, hence, as $\bar{v}$ is a constant times C,

$$nC_B = n_A C_A+n_B C_B \tag{2.10}$$

Combining equations (2.9) and (2.10), it follows that

$$n_A C_A = n_B C_B = \tfrac{1}{2}nC_B \tag{2.11}$$

From the well-known result in the kinetic theory of gases, the pressure p is given by

$$p = \tfrac{1}{3}nmC_B^2 \tag{2.12}$$

Hence, from equation (2.8),

$$F = pa\left(\frac{n_A C_A^2}{nC_B^2}+\frac{n_B C_B^2}{nC_B^2}-1\right)$$

Substituting from equation (2.11) gives

$$F = \tfrac{1}{2}pa\left(\frac{C_A}{C_B}-1\right) \tag{2.13}$$

Equation (1.1) gives $p=nkT$.

Reference to equation (2.12) shows that the root mean square velocity, C, for a given gas, is proportional to $\sqrt{T}$. Hence, equation (2.13) becomes

$$F = \tfrac{1}{2}pa[\sqrt{(T_1/T)}-1] \tag{2.14}$$

The angle of deflection θ of the vane B is decided by the couple exerted on it by the force F balanced against the restoring couple due to the suspension fibre that supports B. This deflection is readily measured by, for example, attaching a light mirror to the vane and using a lamp and scale. This deflection is then a measure of the pressure p *independent* of the nature of the gas molecule.

In the above theory, it is assumed that molecules arriving at the vane B or the plate A acquire the temperatures of these surfaces on impact. In fact, an accommodation coefficient α is involved, decided by

$$\alpha = \frac{T_r - T_i}{T_s - T_i}$$

where T_i is the temperature of the incident molecules at a surface at temperature T_s, and T_r is the temperature of the re-emitted molecules. If $\alpha = 1$, $T_r = T_s$ and the assumption is correct. It may be shown (*see* for example, Leck [60]) that this accommodation coefficient does not change the equation (2.14) even if α is less than unity (normally the case), provided its values at the heated plate A, the vane B, and at the inner walls of the gauge container are the same for all the gases used. This is not easy to arrange in practice, so variations of α with the nature of the gas do cause small variations of the gauge calibration.

A design due to Dumond and Pickels [61] (Fig. 49) is representative of the practice to introduce a fairly robust gauge for general purposes. The aluminium vane is in the form of a rectangular frame $2{\cdot}5 \times 10^{-3}$ cm thick, the inner and outer edges of this frame being folded over to ensure rigidity. This vane is suspended about a vertical axle of aluminium wire of diameter 3×10^{-2} cm, which passes through slits in the frame. A 1 cm diameter galvanometer mirror attached to this axle permits the angle through which the vane turns to be recorded by the use of a lamp and scale. The suspension wire from the steel taper plug at the head of the instrument is of tungsten. The taper plug acts as a torsion head to enable the angle of the vane to be set; after such adjustment, wax is coated over the join of this plug to the top plate to ensure freedom from leakage. The suspension is centralized and prevented from swaying by the eyelets shown. Two nichrome heater coils (forming the plate at the higher temperature) are arranged on either side of the aluminium frame, one opposite each vertical section. A brass cylindrical envelope, with a glass window to admit a beam of light, encloses the gauge, around which water-jackets are arranged to ensure a uniform wall temperature. Helmholtz coils, or a permanent magnet suitably disposed outside the instrument, are used to provide electromagnetic damping of the vane.

This instrument cannot be baked to degas it. Its range is from 10^{-6}

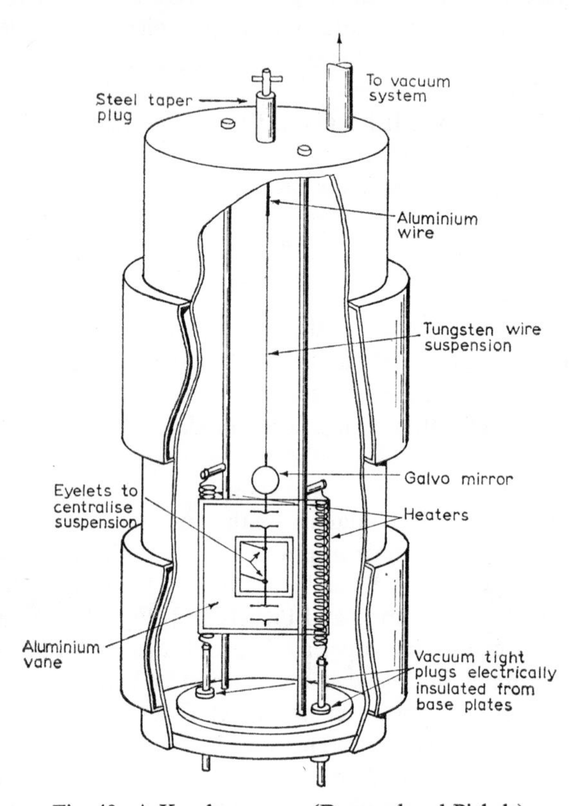

Fig. 49. A Knudsen gauge (Dumond and Pickels).

to 10^{-3} torr. Glass envelope, bakeable, Knudsen gauges have also been constructed, but it is difficult to ensure a minimum pressure reading below 10^{-6} torr because of mechanical vibration causing small vane oscillations.

2.6. *Cold-cathode Ionization Gauges*

When electrons pass through a gas, they will ionize it if their energies exceed the ionization energy of the gas in question. These ionization energies extend from a minimum at 3·9 eV for caesium, through 14·5 eV for nitrogen, up to a maximum for helium, for which the first ionization energy is 24·56 eV. The electrons colliding with gas molecules will produce positive and negative ions. The probability of ionization, decided by the number of ion pairs produced per cm of path of the electron at a given pressure, will vary considerably with the electron energy (a maximum at about 100 eV) and also with the nature of the gas. For a given gas and electron energy, the number of ion pairs produced will also depend on the pressure. If, therefore, an

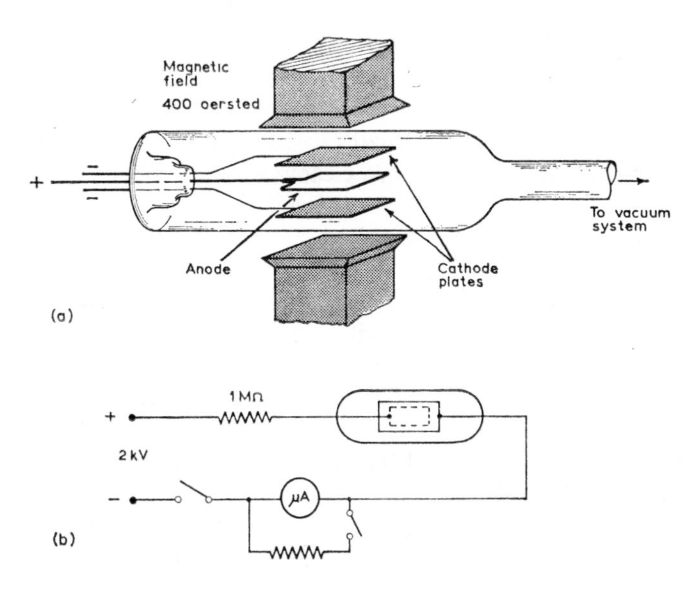

Fig. 50. The Penning gauge.

ionization current is produced in a gas on application of a high voltage between electrodes, the magnitude of this current depends on the gas pressure. As shown in section 1.14, however, it is not possible to sustain sufficient ionization of the gas at low pressures unless the electron paths are greatly increased, which is done by causing them to spiral and oscillate in a discharge confined within a magnetic field. In 1937, Penning [62] introduced an ionization gauge in which a pair of cathode plates is arranged on either side of an anode ring within a glass envelope (Fig. 50*a*), and an axial magnetic field strength of about 400 oersted is maintained to confine the discharge. A significant ionization current, down to a pressure of 10^{-5} torr, is then recordable in a simple design using the series circuit shown in Fig. 50(*b*), with a power-pack H.T. supply of 2,000 volt d.c. and a permanent magnet to maintain the magnetic field. The supply is conveniently a voltage-doubler rectifier unit working from the a.c. mains, the output being smoothed by condensers and chokes to give 2,000 volt d.c. at 20 mA. The ionization current is large enough to be recorded directly by a microammeter supplied with a shunt which is switched in for the higher currents at higher pressures. Alternatively, a tuning glow lamp, such as is used in a radio receiver, will serve as a visual guide to the pressure.

As explained in section 1.14, electrons produced initially in the discharge, and on positive ion impact at the cathode, will travel in tight spirals to and fro about the anode ring, before eventual collection at the anode. In the very long paths described (several hundred times greater than the anode-cathode clearance), the probability of electrons ionizing the gas molecules is significant even at low pressures and corresponding long m.f.p.'s.

The **Penning gauge** is sometimes known as a Philips gauge, because Penning worked at Philips Gloeilampenfabrieken in Holland. A typical calibration curve for air is given in Fig. 51. This calibration will depend on the nature of the gas. Normal practice is therefore to calibrate the gauge for each gas concerned against a McLeod gauge, but recently calibration against a standard orifice (section 3.9) has been preferred. The maximum of the pressure range is at about 5×10^{-3} torr.

Many vacuum equipment manufacturers market Penning gauges, which are frequently used in industrial and laboratory practice for routine, instantaneous pressure measurement obtained by a mechanical/vapour pump combination over the range from 10^{-6} to 5×10^{-3} torr. Both glass and metal envelope types are made. The latter has conveniently the stainless steel, aluminium, or nickel-plated copper envelope with flattened sides, at earth potential, to act as the cathodes; there is an inserted, isolated, anode frame. Though a valuable indicator of pressure, the readings provided should be accepted with reserve. Apart from the fact that the calibration depends on the nature of the gas, significant errors are caused because the gauge pumps gas as the positive ions become embedded in the cathodes, and also because the cathode metal is sputtered, especially at the higher pressures of 10^{-3} torr or more. To minimize error due to this pumping, the gauge must be connected to the system by a short, wide tube, preferably of at least 1 inch bore diameter. Errors are also caused because polyatomic molecules, e.g. of pump oil vapour, in the gauge dissociate in the discharge. Sometimes difficulty with initiating the discharge is experienced at low pressures. To minimize this, a small, sharp point is often attached inside the cathode to ensure production of electrons by field emission. The electrodes of a glass envelope gauge are usually made of nickel, but Penning employed zirconium

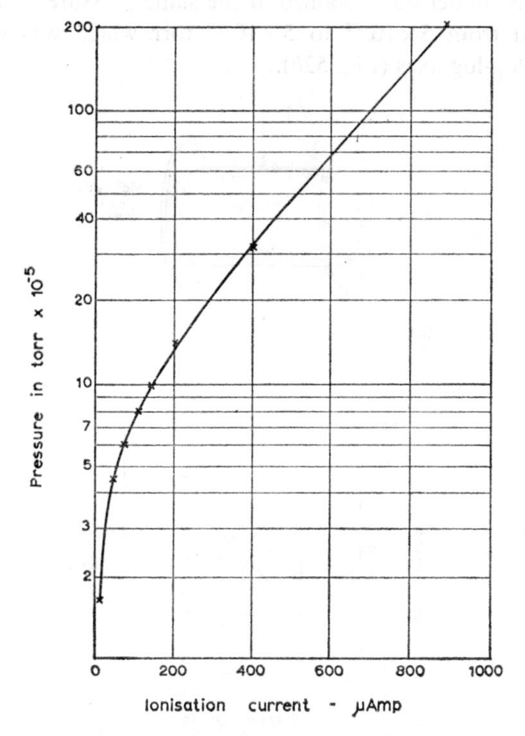

Fig. 51. Ionization current against pressure curve for a Penning gauge containing dry air.

cathodes on the basis that this metal has a low work function, so gives a copious supply of electrons in the discharge, and a low sputtering rate. These gauges cannot be as readily degassed as the hot-cathode ionization gauges (section 2.7); such degassing is best achieved by baking and eddy-current heating *in vacuo*. The gauge should not be run for longer than needed to record the pressure, so that excessive sputtering and pumping are avoided.

A gauge of the type shown in Fig. 50 will give an ionization current of about 10 μA at 10^{-5} torr. It will tend to suffer current fluctuations at a given pressure, and sometimes irreproducible calibration, because of a tendency to ionic oscillation. Penning and Nienhuis [63] reduced these difficulties by a design having a cylindrical instead of a frame anode, so that the discharge was virtually totally enclosed by the electrodes (Fig. 52*a*). They also used a higher magnetic field strength of about 1000 oersted. About 10 times the ionization current in the early model was obtained at the same pressure in air, with a calibration from 5×10^{-7} to 5×10^{-3} torr which was practically linear on log-log axes (Fig. 52*b*).

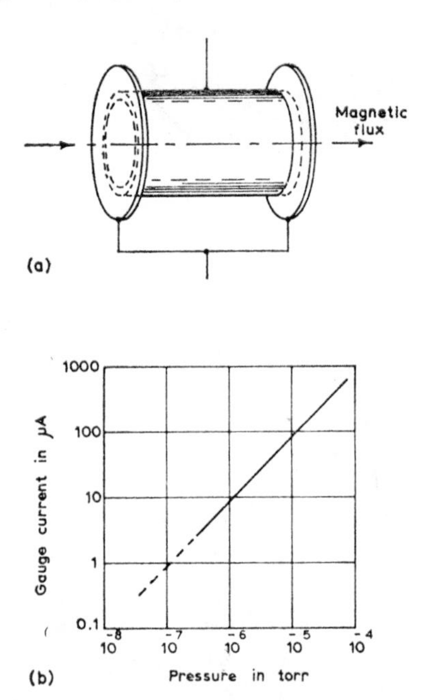

Fig. 52. The Penning-Nienhuis cold-cathode ionization gauge.

In recent developments of the cold-cathode ionization gauge, perhaps the most outstanding work is that of Redhead [64], who developed the inverted magnetron gauge and the magnetron cold-cathode gauge for use in the ultra-high vacuum region down to and even below 10^{-12} torr. Of the two, the inverted magnetron gauge has been developed most, and commercial models are now available. The term 'inverted magnetron' arises from its similarity to the magnetron valve, with a central axial electrode surrounded by a separate cylindrical electrode; but here the central one is the anode and the outer one the cathode, the reverse to the case of the magnetron itself. The term **Redhead gauge** is also frequently used.

In extending the range of the cold-cathode gauge to far below 10^{-7} torr (the limit of the Penning-Nienhuis type), Redhead appreciated that the limit to the conventional type was set by field emission electron current from the cathode. This was obtained on attempting to use a p.d. considerably higher than 2 kV to ensure satisfactory initiation of the discharge at very low pressures. This field emission current is independent of the pressure and so sets the lower limit. Further, very efficient trapping of electrons executing very long paths was essential to obtain recordable currents in the ultra-high vacuum range. He therefore constructed a design in a glass envelope with a central axial anode wire of tungsten, 0·25 mm diameter, surrounded by *two* cylindrical box-shaped cathodes. The inner one of these cathodes was 30 mm in diameter and 15 mm long, made of polished Nichrome V, and earthed through the d.c. amplifier used to record the ionization current. The surrounding isolated auxiliary cathode, also of Nichrome V, was directly earthed, i.e. was not connected to the d.c. amplifier. This auxiliary cathode was also provided with short cylindrical shields each 6 mm in diameter, projecting 2 mm inside the ion collector cathode, from which they were isolated. A positive potential of 6 kV with respect to earth was maintained on the anode, and the axially-directed, uniform magnetic field was 2,000 oersted approximately (Fig. 53*a*).

Apart from acting as an electrostatic shield against external disturbances to the ion collector cathode, the auxiliary cathode, with its short, projecting cylinders spaced 3 mm from the anode wire, provided the initial field emission electrons needed to start the discharge. But this field emission current was not recorded by the d.c. amplifier. These short cylinders also protected the end plates of the ion collector cathode from the electric field, so preventing field emission from this cathode to the anode wire.

A calibration curve for this gauge obtained by Redhead, with a magnetic flux density of 2,060 gauss and an anode potential of 6 kV, is shown for dry air in Fig. 53(*b*). The ion-collector current, I_c, and the pressure, p, are related by the equation

$$I_c = kp^n$$

where k is a constant of the gauge depending on the nature of the gas, and n is a constant for the gauge of value between 1·1 and 1·4.

The gauge is useful over the pressure range from 10^{-3} to 10^{-12} torr. At the low end of this range, the discharge took as long as 10 min to strike after application of the anode voltage, but striking apparently always took place.

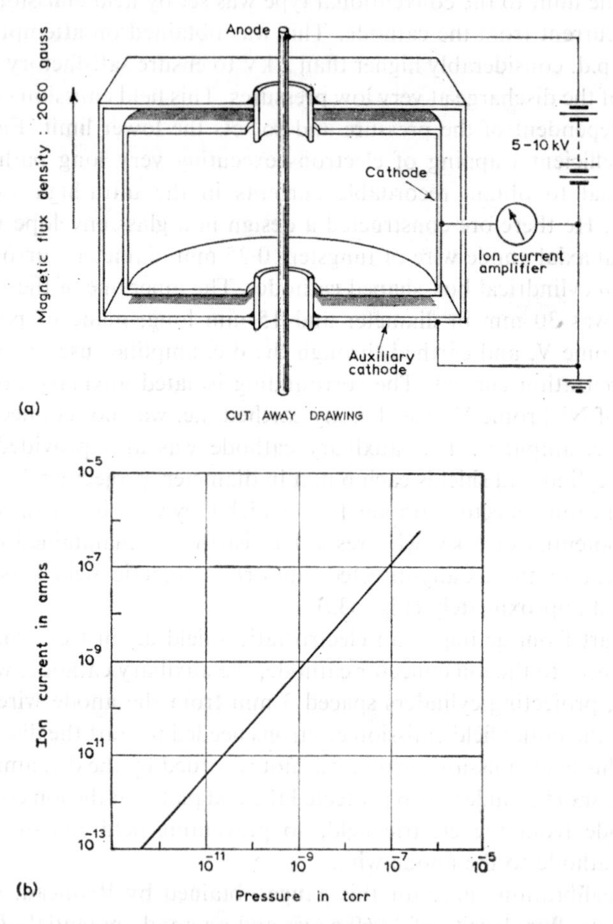

Fig. 53. The inverted magnetron cold-cathode ionization gauge.

For dry air, this gauge provides an ionization current of about 1 mA per millitorr. To record pressures down to 10^{-12} torr, therefore, a good quality, stable, d.c. amplifier or vibrating-reed electrometer is needed to record the current.

2.7. *Hot-cathode Ionization Gauges*

The most widely used vacuum gauge for pressure measurement below 10^{-4} torr is the hot-cathode ionization gauge. It was introduced first by Buckley [65] in 1916, and developed in 1921 by Dushman and Found [66]. In its older, conventional form, this gauge is essentially similar to a triode valve. Electrons from a heated filament, in an envelope at the pressure to be determined, are accelerated towards an electrode maintained at a positive potential of 100 to 200 volt, well above the ionization potentials of the gases. Many of these electrons collide with the residual gas molecules in the envelope to form positive ions which are collected on a third negative electrode. The posi-

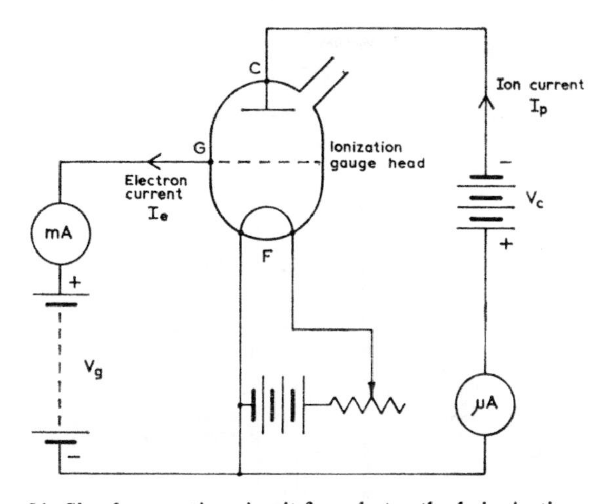

Fig. 54. Simple operating circuit for a hot-cathode ionization gauge.

tive ion current to this negative electrode varies with the gas pressure. With the conventional triode valve geometry, the usual practice is to make the grid positive and the 'anode' negative with respect to the filament. The simple, basic, operating circuit is therefore as in Fig. 54. The electron current to the grid, I_e, is recorded by the milliammeter, mA, and maintained at, say, 1 mA; this value being achieved by adjusting the filament current to obtain the appropriate filament temperature and emission. The grid is at a constant positive potential V_g of, say, 150 volt. The electrons reach the grid with energies of 150 eV. Many of them traverse the spaces between the grid wires, are then retarded by the negative 'anode' (best called the ion collector C) back towards the grid, and tend to oscillate to and

fro about the grid a few times before being collected at the grid wires
The electrons ionize the gas in the region about the grid, and the posi
tive ions formed go to the ion collector which is maintained at V
equal to, say, -25 volt with respect to the filament. These positiv
ions reaching the ion collector provide an ion current I_p, which i
recorded by the microammeter, μA, in series with the supply of nega
tive potential to this electrode. This ion current will vary linearl
with the gas pressure for a given gas, electron current, and potential
on the grid and ion collector. For dry air at a pressure of 1 millitorr
this current, I_p, will be about 20 μA, with I_e equal to 1 mA.

An example of an earlier conventional gauge head (Fig. 55), due t

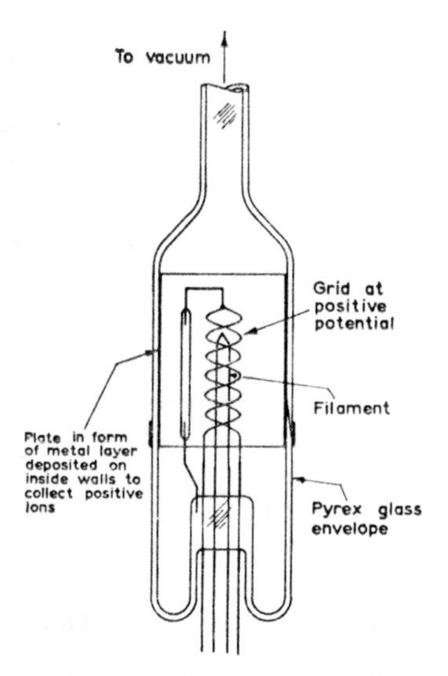

Fig. 55. Hot-cathode ionization gauge head (Morse and Bowie).

Morse and Bowie [67], is made of Pyrex glass in which a hair-pin-shaped tungsten filament is set up as the electron emitter. This filament is surrounded by a grid in the form of a spiral of tungsten wire. The ion collector electrode is a thin, translucent, platinum film deposited over the central region of the inner wall of the Pyrex tube. Electrical contact is made to this film by means of a wire embedded in the glass, with an external contact in the form of a metal ring round the centre of the gauge head. The recommended operating parameters are $+150$ volt on the grid with respect to the filament, -25 volt on the collector, and an electron current to the grid of 5 mA.

This gauge head design can be readily degassed, because the grid may be directly heated on the passage of current by connecting a 6 volt supply to the appropriate pinch leads; moreover, the envelope and the ion collector electrode may be baked in an oven or heated with a blow-pipe flame.

For a given gas in the gauge, with given operating potentials, V_c on the ion collector and V_g on the grid, the positive ion current I_p is directly proportional to the gas pressure over the range from 10^{-8} to 10^{-3} torr approximately and to the electron current I_e. Thus,

$$I_p = SpI_e$$

$$\therefore \; S = I_p/pI_e \tag{2.15}$$

where S is the sensitivity of the gauge. This will depend markedly on the nature of the gas because of variation, from one gas to another, of the probability of ionisation by electrons of a given energy. The sensitivity is usually given for dry air or dry nitrogen. S is quoted with I_p in μA, I_e at 1 mA, and for a pressure of 1 millitorr, i.e. in μA per mA per millitorr; or with I_p and I_e in amp, at 1 torr, i.e. in amp per amp per torr, equivalent to per torr. Note:

$$S \text{ in } \mu\text{A mA}^{-1}\text{ m}T^{-1} = S \text{ in torr}^{-1}$$

A gauge with $S=20$ per torr, for dry nitrogen, will give a positive ion current I_p of 2×10^{-10} amp, for an electron current of 1 mA at a pressure of 10^{-8} torr. To measure these small ion currents, a d.c. amplifier or a vibrating-reed electrometer is required, because the limit of the usual box galvanometer is about 10^{-7} amp for a deflection of 10 mm. Using a d.c. amplifier, the basic operating circuit is shown in Fig. 56. Here the electrode C is connected to the d.c. amplifier, which is conveniently earthed. Hence the ion collector C is at zero potential before an ion current is established. Consequently, the mean potential of the filament F must be made positive with respect to earth. This is achieved by a resistor R between the negative side of the filament and earth. As R carries the electron current I_e, there is a p.d. of I_eR maintained across it (where R is the value of the resistor). As I_e is usually 1 mA, to make the p.d. across R equal to 25 volt, R is selected to be 25 kΩ.

Very low pressure ionization gauges. The lowest pressure which can be recorded by a conventional hot-cathode ionization gauge is about 10^{-8} torr. At this pressure, the positive ion current is about 2×10^{-10} amp, as shown previously. There is no difficulty in measuring such currents with a calibrated d.c. amplifier; indeed, measurement down to 10^{-14} amp or lower is expected practice with good quality d.c amplifiers and vibrating-reed electrometers, Why, therefore, will the gauge not indicate much lower pressures? Nottingham [68] suggested and it has since been amply confirmed, that the chief limitation is se by soft X-rays produced at the grid by electrons striking it with energies of V_g eV, with V_g about 150 volt. These soft X-rays wil irradiate the ion collector electrode C; in doing so they will releas electrons from C. The electrons leaving C form a residual current which is added to the current I_p due to positive ions arriving at C. The total current is therefore I_p+I_x, I_x being the X-ray photoelectri

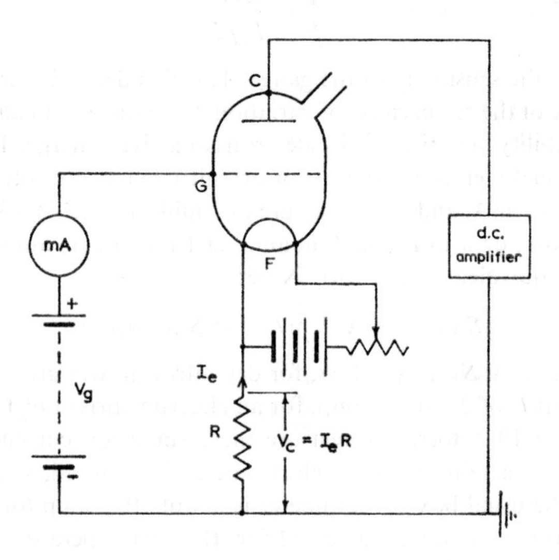

Fig. 56. Basic operating circuit of a hot-cathode ionization gauge with d.c amplifier.

current. Whereas I_p varies with the pressure, I_x does not. In the con ventional gauge, I_x is equal to I_p at a pressure of about 10^{-8} torr Hence, below this pressure, the recorded current to the ion collecto electrode C becomes largely independent of the pressure.

Soon after Nottingham's suggestion, Bayard and Alpert [69] intro duced a design of hot-cathode ionization gauge head in which thi residual constant current, I_x, due to X-rays, was reduced by a facto of from 100 to 1,000 or more. The gauge they introduced was the able to record pressures down to 10^{-10} torr and below. The ingeniou solution was to reverse the positions of the filament and the ion col lector: the filament was erected outside the cylindrical grid and th ion collector was made a thin wire along the central axis inside the grid. In the design of Fig. 57, two hair-pin tungsten filaments, FF, are present, one acting as a spare. The thin tungsten wire, ion collector electrode has a diameter of 0·1 mm or less. X-rays are still produced at the grid on impact of the electrons from the filament, but their interception by the ion collector wire is now 100 to 1,000 times less than it was by the conventional, outer, cylindrical ion collector.

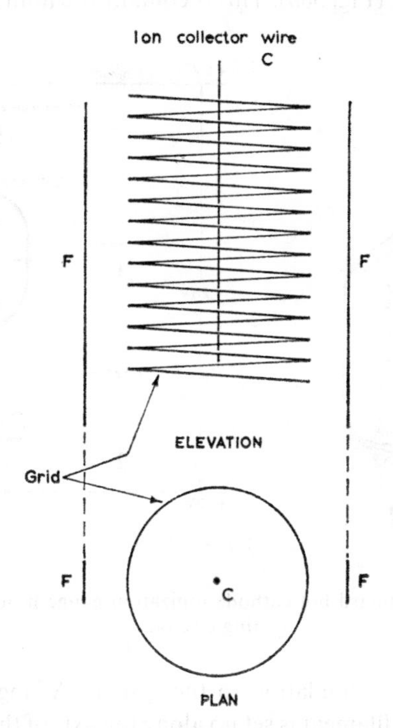

Fig. 57. Principle of the Bayard-Alpert hot-cathode ionization gauge.

So the residual X-ray current I_x is reduced correspondingly, being comparable with I_p at pressures now as low as 10^{-10} torr or less. This type of gauge still has a sensitivity S comparable with that of the conventional design, because the ionization region is chiefly within the considerable volume just inside the cylindrical grid, but it now has a range of response from 10^{-3} to 10^{-10} torr or below. Indeed, by far the majority of hot-cathode ionization gauges are of the Bayard-Alpert type, as they operate as satisfactorily as the conventional pattern with the advantage of an extended pressure range. Furthermore, they are easier to degas because of the absence of the large cylindrical ion collector which is replaced by a thin wire. The technique of obtaining ultra-high vacua may be said to have begun in 1950, with the introduction of this **Bayard-Alpert gauge** (B.A.G.), because it provided for the first time a satisfactory means of recording pressures below 10^{-8} torr.

A good model of commercial Bayard-Alpert gauge is the Mullard Ltd. type IOG-12 (Fig. 58*a*). This is contained within a glass envelope

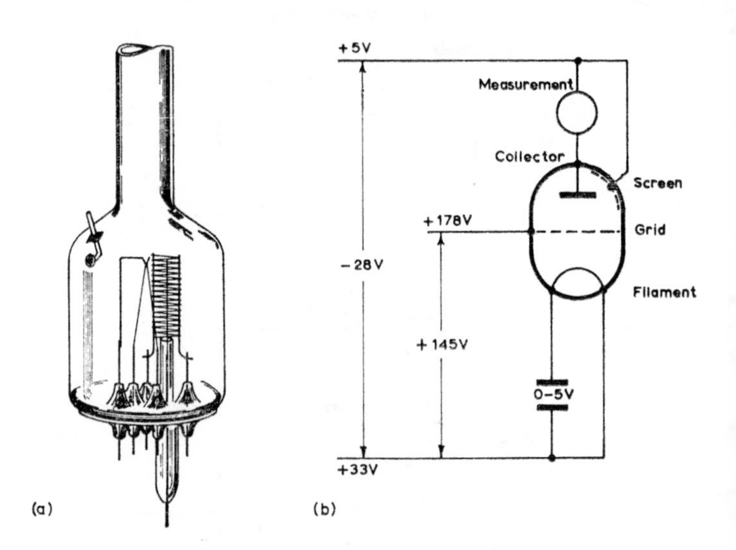

Fig. 58. (*a*) The Mullard hot-cathode ionization gauge head. (*b*) Basic operating circuit.

for attachment by tubulation to the system. A single, inverted, V-shaped, tungsten filament is set up along the axis of the glass envelope to avoid asymmetrical heating and degassing of the wall, brought about in the original Bayard-Alpert design with its off-axial heated filament near one section of the wall. Furthermore, the inner wall of the glass envelope is coated with tin oxide. This forms a conducting layer to which connection can be made by a lead-in through a glass-to-metal seal. By maintaining this layer at the ion-collector potential (Fig. 58*b*), two advantages are gained; excess primary electrons are attracted directly to the walls, rather than oscillating around the collector, ensuring stable measurements; and possible bistable operation of the gauge is avoided. Carter and Leck [70] have shown that the glass surface of the normal B.A.G. envelope can become stabilized at a potential equal to that of either the cathode or the positive grid; the second case particularly occurs at grid potentials above +250 volt. Changes from one stable state to the other are unlikely at grid potentials of +150 volt, but, if they occur, considerable alterations result in the gauge characteristics. Moreover, it is preferable to ensure a definite glass surface potential, rather than one decided by the ratio of the secondary to the primary electrons at this surface.

The pressure range of the Mullard IOG-12 is 10^{-3} to 10^{-11} torr, with a sensitivity S for dry nitrogen of 12 per torr. Typical operating potentials are shown in Fig. 58(*b*). The calibration curve of positive ion current against pressure, with an electron current of 1 mA, is linear over the range 10^{-3} to 10^{-11} torr.

In the operation of a hot-cathode ionization gauge, it is essential to degas thoroughly the envelope and the electrodes. The degassing is first achieved by bake-out of the gauge, when under vacuum at 10^{-6} torr or less, by heating in an oven to 450°C (assuming a borosilicate-glass envelope). Degassing of the electrodes can be by eddy-current heating or by the passage of current directly through the grid of a design such as that of Morse and Bowie. The usual and best practice, however, is to degas the electrodes by electron bombardment. To do this, the filament current, and therefore emission, is increased above normal values (filament voltage is 8 volt with a filament current of 1·9 amp, for degassing the IOG-12, whereas this current is only 1·45 amp at a voltage of 4·4 volt to obtain $I_e = 1$ mA), and a positive potential of 400 to 600 volt is applied to the grid and ion collector strapped together (the conducting layer of the IOG-12 is left floating in potential during degassing). The wattage dissipated in these electrodes is then the product of I_e, the electron current in amp, and V_g the positive grid voltage. A dissipation of about 40 watt is demanded in the grid of a B.A.G. to heat it to about 800°C, so that it glows red. The IOG-12 gauge, for example, is degassed at $V_g = 550$ volt and $I_e =$ 75 mA, giving a dissipation of $550 \times 0{\cdot}075 = 41$ watt. This degassing should be continued for 30 to 60 min, at the lowest attainable pressure, when first setting up the gauge. Cleanliness of the grid is particularly important in recording ultra-high vacua, and such degassing should be undertaken whenever it is suspected that the electrodes have sorbed gas.

A hot-cathode ionization gauge exhibits a **chemical** and an **electrical pumping action**. The former is due to gas reacting at the hot filament; this is particularly prevalent with oxygen, due to chemisorption and chemical action, and with hydrogen, because it dissociates partly to atomic hydrogen when coming into contact with a filament at high temperature, and atomic hydrogen is readily sorbed at the gauge wall. The electrical pumping is due to positive ions which are sorbed on reaching the ion collector and also the glass wall at or near earth potential. Thus, a positive ion with an energy, say, 150 eV, created near the grid, and travelling to the ion collector or the wall, will have a distinct probability of becoming buried in the surface. Again, and in addition to the chemical pumping, the filament of a hot-cathode ionization gauge, operated in a vacuum created by an oil diffusion pump, without efficient trapping against back-streaming, will be likely to dissociate oil molecules and react in an uncertain way with the decomposition products.

A gauge head which pumps gas, and which may also evolve and dissociate gas, will be at a different pressure from that prevailing in the system, to which it is connected by tubulation. This pressure difference will be the more pronounced the lower the conductance to gas flow of this tubulation. As emphasized by Blears, [71] it is therefore necessary to ensure that the connecting tubulation used is short and of as large a diameter as possible, to minimize pressure gradients between the gauge head and the system. In general, this tubulation should have a diameter exceeding 1 inch.

In this connection, it is good practice, if possible, to immerse the gauge electrodes directly into the chamber in which the pressure is to be recorded. For this purpose, several manufacturers provide nude gauge heads, an example of which is shown in Fig. 59. For the same electrode geometry, the removal of the envelope from the gauge changes the sensitivity and probably reduces residual current effects resulting from desorption of ions.

The thoroughly degassed gauge head attached to an ultra-high vacuum system can give misleading results near the ultimate pressure, where the pumping speed of the system itself is small. This is because of the pumping action of the gauge itself. Considering the electrical pumping, suppose I_p is the positive ion current to the ion collector and I_w the positive ion current to the walls. Assume that all the ions which reach the walls and collector are retained and each ion has a single positive charge of numerical value e, the electronic charge. The number of ions sorbed per sec is $(I_p+I_w)/e$, where I_p and I_w are in amp, and e is in coulomb. Both I_p and I_w increase in proportion to the electron current I_e to the grid. To reduce the electrical pumping speed, I_e is therefore reduced in magnitude. Generally, I_e is set at 100 μA, or as low as 10 μA, to ensure that electrical pumping is negligibly small. As this reduction of I_e is brought about by reduction of the filament temperature, chemical pumping and other reactions at the filament are also reduced.

Further reactions of the hot tungsten filament with the gas in the gauge head can frequently be troublesome. For example, commercial-grade tungsten wire contains about 0·01% by weight of carbon. If operated as an electron emitter at, say, 2,000°C, and at pressures up to 10^{-4} torr, the partial pressure of the oxygen present will drop rapidly because of the formation of carbon monoxide (CO) and

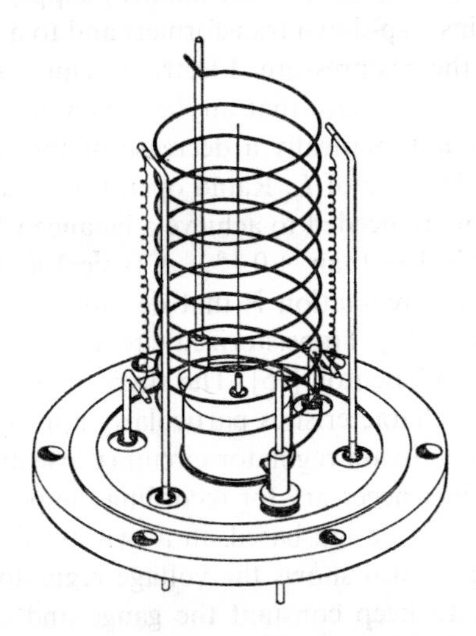

Fig. 59. A nude hot-cathode ionization gauge head (based on gauge of Vactronic Lab. Equipment, Inc.).

carbon dioxide (CO_2) at the filament. The partial pressures of these gases will rise.

The remedy is to use a thermionic emitter of lower work function than tungsten, so that comparable emission per unit surface area can be obtained at considerably lower temperatures. Oxide-coated filament emitters have been used, but unfortunately tend to become poisoned in the presence of some reactive gases, particularly oxygen and hydrocarbons. Chemically-inert rhenium is useful here as an alternative to tungsten. The best practice is probably to use lanthanum hexaboride (LaB_6)-coated rhenium, giving an emission of 10^{-1} amp per sq cm at a temperature of 1,250°C; whereas pure tungsten has to be heated to 2,200°C to give this emission.

This ability to use a much lower filament operating temperature also gives longer filament life. Thoria-coated iridium or rhodium filaments have also been used successfully (Weinreich and Bleecher [72]), but at present most commercial gauge heads still have tungsten filaments.

A provision, which is virtually a necessity, to ensure reliable operation of a hot-cathode ionization gauge without frustration, is an **emission regulator**, whereby the electron current I_e to the grid is kept constant. With fixed values of the positive grid voltage V_g, variations of I_e occur due to variations of the filament supply voltage (usually from an a.c. mains step-down transformer) and to a small extent due to variations of the gas pressure. Electronic emission regulator circuits are designed to ensure that any tendency for I_e to increase is used to monitor automatically a decrease of the filament current; vice-versa for a decrease of I_e. Rapid operation and adequate sensitivity and control are needed to achieve a balance whereby I_e is kept constant to within less than $\pm 0{\cdot}1\%$. A typical and excellent valve circuit for emission regulation is that developed by Steckelmacher and Van der Meer [73]. Transistorized circuits have been developed by Holmes [74] and Benton [75]. The thyristor is a semiconductor device which has characteristics particularly appropriate to the development of an emission regulator circuit (Close and Hodges [76]).

The d.c. amplifier necessary for recording the positive ion current can be of the simple design based on a twin-triode valve shown in Fig. 60. This figure also shows the voltage regulator tubes, V_2 and V_3, which serve to keep constant the gauge and circuit operating potentials. A better d.c. amplifier due to Allenden [77], which has been found particularly good for recording currents down to 10^{-12} amp or below, is based on ME1403 electrometer valves. A simple, easily-constructed d.c. amplifier employing an ME 1401 electrometer valve with battery operation, which has been found very convenient for college experiments on ionization gauges, is described by Yarwood and Close [78]. The new metal oxide, semiconductor transistors (MOST) are very attractive for d.c. amplification (Close [79]).

The range of the Bayard-Alpert ionization gauge is from 10^{-3} to 10^{-10} torr or below. Redhead [80] has extended the lower end of this range by a factor of 10 or more by introducing a **modulated Bayard-Alpert gauge** (Fig. 61). The normal B.A.G. head construction is used with the addition of a modulator electrode in the form of a straight wire set up parallel to the ion collector and just inside the cylindrical grid. The potential on this electrode is switched between earth and

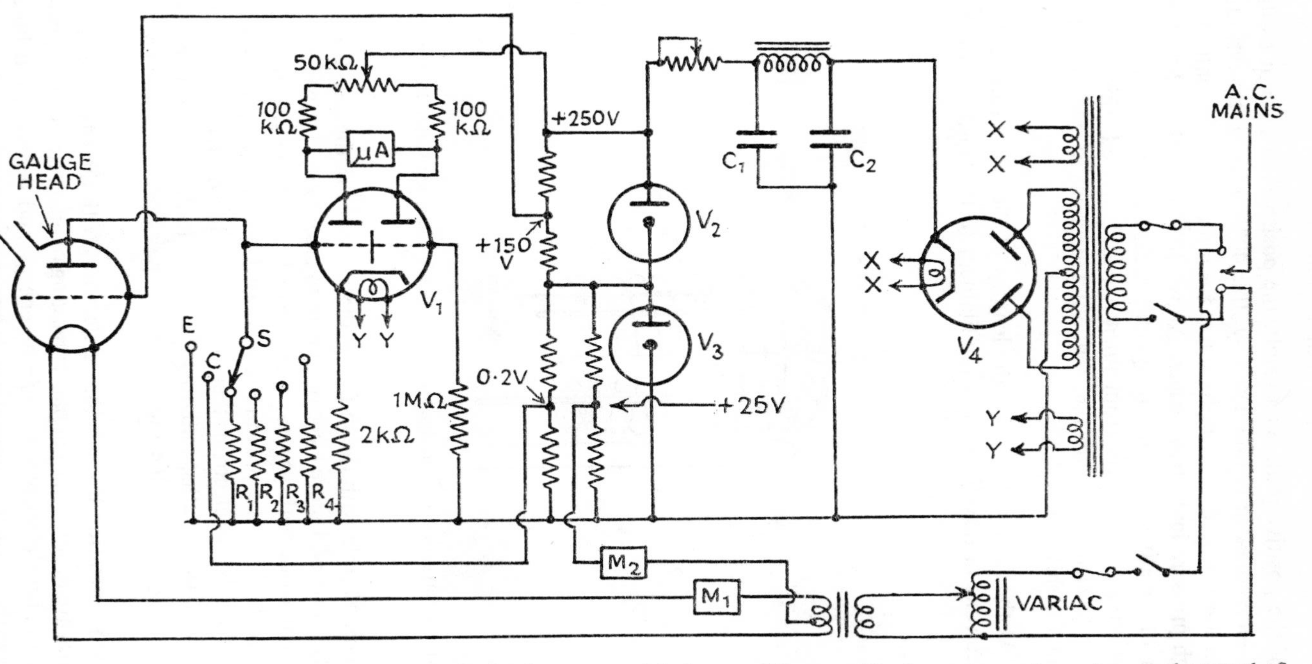

Fig. 60. An operating circuit for a hot-cathode ionization gauge with d.c. amplifier, stabilized power supply, and manual control of the emission current.

that of the positive grid. As a result, the positive ion current to the earthed ion collector is modulated by 30 to 40%, whereas the residual current due primarily to X-radiation of the ion collector is unaffected.

High pressure ionization gauges. Above 10^{-3} torr, the calibration graph of positive ion current against pressure for a hot-cathode ionization gauge departs from linearity, and, indeed, this current tends to level off to a constant value independent of pressure. The reasons for this follow.

(*a*) At the higher pressures, the lengths of the paths executed by the electrons in travelling from the filament to the positive grid be-

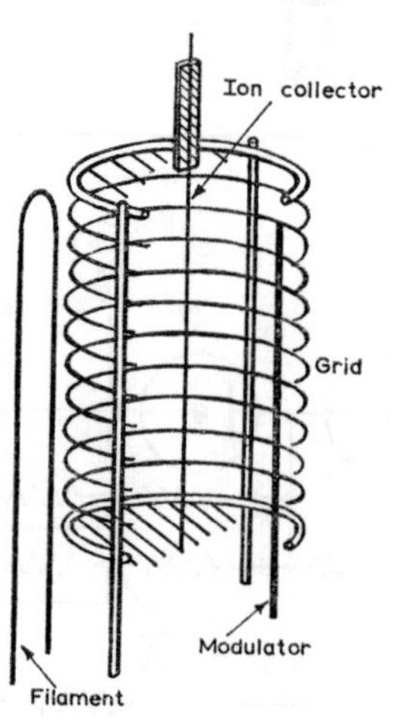

Fig. 61. A Bayard-Alpert gauge with modulator electrode (after Redhead).

come comparable with the m.f.p. of the electrons in the gas. Consequently, more and more electrons begin to lose energy by making inelastic collisions with gas molecules, which result in excitation but not ionization, as the pressure is increased. Significant fractions of the total number of electrons therefore have reduced energies towards the ends of their paths, and so have lower ionization probability, which is expressed as the number of ion pairs produced per cm of path.

(*b*) On ionization by collision of electrons with the gas molecules, ion pairs are produced, each pair consisting of an electron and a positive ion. These electrons are of low energy and do not contribute significantly to further ionization. However, they will travel to the positive grid. The electron current, I_e, to the grid is maintained constant during measurement. If a considerable fraction of this current is due to these low energy electrons and not to those directly emitted from the hot filament, the sensitivity of the gauge will appear to fall off. This effect becomes more pronounced as more ion pairs are created, at the higher pressures.

(*c*) A positive-ion space charge begins to form in the filament region, and a glow discharge appears at pressures of a few millitorr. Owing to the plasma formed, scattering of positive ions away from the ion collector to the walls and other electrodes begins to become large enough for the ion-collector current to fall off significantly. This may be counteracted to some extent, because loss of ions from the collector, due to axial drift away from the discharge space, is greater at the lower pressures.

Schulz and Phelps [81] have designed an ionization gauge (Fig. 62)

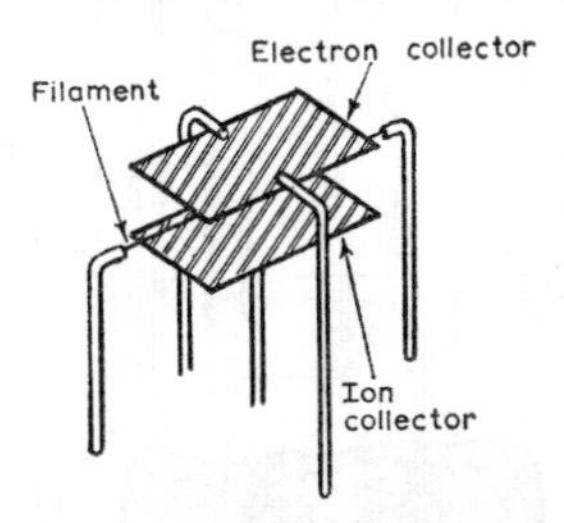

Fig. 62. The Schulz and Phelps hot-cathode ionization gauge head electrodes with a range from 10^{-6} to 0·6 torr.

in which these three factors are minimized in their effect, to the extent that it can be used to record pressures without departure from linear calibration up to 0·6 torr. Their simplest design (Westinghouse model WX4145) consists of a straight iridium filament of diameter 0·125 mm along the axis between two parallel molybdenum electrodes set 0·125 inch apart, each 0·5 by 0·375 inch. One of these is the 'grid', at a positive potential of 60 volt, and acts as the electron collector, and the other is the ion collector at a negative potential of −60 volt. The short electron path without oscillation from the filament to the positive plate at a comparatively low potential ensures that an electron is unlikely to make more than one collision, so reducing greatly the contribution due to factor (*a*), but at the sacrifice of gauge sensitivity, which is only 0·6 torr^{-1} for nitrogen. This provision further limits factor (*b*), to the extent that those electrons resulting from ion pair formation are less than 10% in number of those emitted from the filament, even at a pressure of 0·15 torr. Effective positive ion collection at the higher pressures (factor *c*) is brought about by the provision of an ion collector electrode of large surface area compared with the filament, and at a high negative potential (−60 volt) to ensure parallel plane equipotential surfaces between the electrodes. However, the low sensitivity combined with the necessary large ion

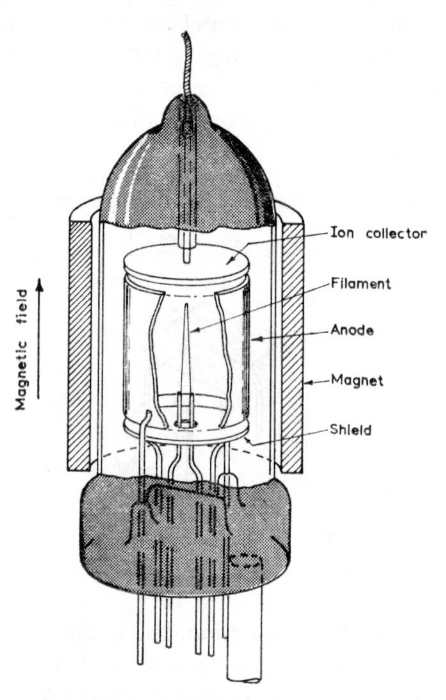

Fig. 63. The Lafferty hot-cathode ionization gauge head.

collector area will mean that the low pressure limit of this gauge is at about 10^{-6} torr, due to the X-ray effect.

This gauge has been used for measuring the pressures of chemically active as well as inert gases in the pressure range from 10^{-6} to 0·6 torr.

Extended electron path gauges. Several attempts have been made to increase the sensitivity of hot-cathode ionization gauges by causing the electrons to execute very long paths from the filament to the electron collector. One of the most successful of these is the **Lafferty gauge** [82], or **hot-cathode magnetron ionization gauge** (Fig. 63). This consists of a hair-pin hot filament along the axis of a cylindrical positive anode with negative end plates, one of which acts as the ion collector and the other as a shield. A magnetic field provided by an external cylindrical magnet maintains a flux density B directed along the axis of the anode. This flux density is large enough to ensure that electrons emitted from the filament spiral around the axis in many orbits before being collected at the positive anode, i.e. the flux density exceeds the magnetron cut-off value. The electron density within the enclosed space between the electrodes is therefore greatly increased by the magnetic field, and, owing to the very long paths they execute, the probability of ionizing the gas by these electrons is considerably larger. The positive ion current is recorded to one or both of the disc-shaped end plates. With a 0·2 mm diameter, tungsten filament, $\frac{3}{4}$ inch long, an anode of $\frac{15}{16}$ inch diameter and $1\frac{1}{8}$ inch long, and a magnetic flux density of 250 gauss, the positive ion current is 25,000 times the value obtained without the magnetic flux, but the electron current is reduced by 50. The ratio of the ion current to the X-ray photocurrent is thus increased $1{\cdot}25 \times 10^{6}$ times, so that the ion current should only equal the X-ray photocurrent at pressures of $2{\cdot}5 \times 10^{-14}$ torr, giving the lower limit of the gauge.

The anode potential was 300 volt, positive with respect to the filament, and the emission current was adjusted by the filament temperature to be 10^{-7} amp, with zero magnetic field. This very low emission avoids the excessive pumping action and liability to unstable operation due to oscillation, prevalent in previous designs.

General considerations. The outstanding advantage of the hot-cathode ionization gauge is its ability to record pressures in the range from 10^{-3} to 10^{-11} torr or below, using the Bayard-Alpert type. It is rivalled only by the cold-cathode gauge in this connection. Indeed, it is the common choice for low pressure measurements and is very widely used.

Its disadvantages are apparent from the foregoing descriptions. From the practical point of view, these may be summed up by stating that any pressure measurement by these gauges is suspect in that it is difficult to know the errors involved. Certainly, any value quoted to more than one significant figure is dubious, even with the best practice. Thus, one might believe a result of, say, 3×10^{-9} torr, but certainly not $3{\cdot}2 \times 10^{-9}$ torr, and even the first quote may really mean that the pressure is anywhere between 10^{-9} and 5×10^{-9} torr!

The disadvantages and problems associated with the use of hot-cathode ionization gauges may be summarized briefly.

(*a*) The calibration depends on the nature of the gas. Calibration against a standard (the McLeod gauge, or best the standard orifice, section 3.9) is essential for every gas used. At the low pressures concerned, the problem is aggravated because the constitution of the residual gases is often unknown, and the presence of small amounts of impurity can markedly affect ionization characteristics of gases. In many cases, a gas analyser in the form of a mass spectrometer or omegatron (section 2.9) is essential. Even then difficulties arise, because switching-on the filament of the gauge or the gas analyser can significantly alter the gas composition at ultra-high vacua.

As a guide to the variation of sensitivity of hot-cathode ionisation gauges with the gas, Leck [83] gives tabulated data. Typical results are from the Edwards High Vacuum Ltd. IG-2 and IG-3 gauges, which have the following sensitivities, *S*, referred to nitrogen as unity.

Gas	N_2	O_2	H_2	CO	CO_2	H_2O	Hg	He	Ne	A
S	1·0	0·8	0·4	1·05	1·37	2·0	2·7	0·14	0·25	1·35

(*b*) The gauge head is a source of gas and also pumps gas. The head must therefore be very thoroughly degassed, maintained clean, operated with low electron currents to the grid, preferably have a low temperature filament, and be separated from the position at which the pressure is to be determined by a path of as large a conductance to gas as possible.

2.8. *Radioactive Ionization Gauges*

A principle of low pressure measurement, first applied to the development of a vacuum gauge by Downing and Mellen [84], is to ionize the residual gas in an envelope by the radiations from a radioactive isotope. Alpha-particles from radium and beta-particles from tritium have both been used. The ionization current measured between two electrodes in the gas is a linear function of pressure over a wide range.

The gauge head due to Downing and Mellen (Fig. 64), known as the **alphatron**, employs a pellet of gold-radium alloy containing 0·2 mg of radium in equilibrium with its disintegration products. This is sealed in a small capsule with a wall thin enough to allow the escape of the alpha-particles but not the escape of radon gas, a daughter product of radium. This capsule is at the base of a box-shaped grid with a central, insulated, cage-type electrode within the gauge housing. A stable d.c. potential of 30 to 40 volt is maintained across the two electrodes, the outer grid being positive, with a d.c. amplifier in series to measure the ionization current. The radium with its very long half-life of 1,622 years emits alpha-particles at a constant rate; these ionize the gas. The ionization current is 2×10^{-10} amp at 1 torr

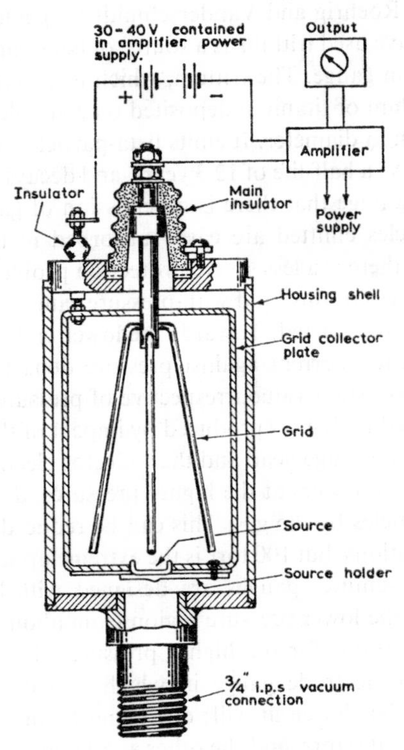

Fig. 64. The alphatron.

for dry air, and is directly proportional to the pressure over the range from 10^{-3} to 100 torr or higher. Unfortunately, the low pressure limit is at 10^{-3} torr because of the difficulty of measuring accurately currents less than 2×10^{-13} amp, but later gauges have been developed using more powerful sources of radium in which 10^{-5} torr can be recorded. Moreover, designs able to record up to 1,000 torr have been made. The calibration depends on the nature of the gas. The linear calibration of ionization current against pressure for a given gas is a considerable advantage, as also is the fact that there is no filament to burn out or react with gas. The difficulty of measuring the lower pressures, the precautions demanded to avoid exposure of personnel to the gamma radiation from the radium, and their rather high cost have precluded their general use in the laboratory. An interesting special application has been to install the alphatron in rockets, to measure the atmospheric pressure at high altitudes.

Vacca, [85], Roehrig and Vanderschmidt [86], and also Blanc and Dagnac [87] have used tritium as a source of beta-particles in a radioactive ionization gauge. The tritium, which is a gas (hydrogen 3), is absorbed in a film of titanium deposited on a stainless steel or silver disc about 3 cm in diameter. It emits beta-particles with a maximum energy of 18 keV, a half-life of 12·3 years, and decays to stable helium 3 with the advantage that there is no emission of gamma radiation. The beta-particles emitted are easily absorbed in the walls of the gauge head, so there is a less severe protection problem than with the alphatron. Gauges of this kind with pressure ranges from 10^{-4} to 10 torr have been constructed. Towards the lower end of the range, the curve of ionization current against pressure departs from linearity and tends to a constant value irrespective of pressure because of the electrons released by X-rays produced by impact of the beta-particles on the walls of the gauge head and the collector electrode. Departure from linearity also occurs at the higher pressures, due to absorption of the beta-particles by the gas. This can be reduced by using small electrode separations, but 100 torr is the extreme upper limit possible. In general, beta-emitter gauges are designed with larger electrode separations for the lower pressure region from about 10^{-4} to 1 torr, and small separations for the higher pressures. Blanc and Dagnac describe a three-electrode gauge in which two different electrode separations can be chosen at will; one being 5 mm for the pressure range from 10 to 100 torr, and the other at 50 mm for the range 10^{-4} to 10 torr.

2.9. *Partial Pressure Measurement: Gas Analysers*

In many cases of low pressure measurement, especially in the region below 10^{-5} torr, it is not enough to know the total pressure; the partial pressures of the various constituent gases, and what these gases are must be known. Thus, a process may be such that it would be ruined on being undertaken at a total pressure of 10^{-8} torr, if the residual gas were chemically active, say oxygen, whereas a total pressure of 10^{-6} torr would suffice if the partial pressures of the active gases present were, say, below 10^{-9} torr.

There are several types of gas analysers and many hundreds of papers have been written about them. Many are able to determine the residual gases present but without recording satisfactorily their partial pressures; some can be calibrated to record partial pressures with considerable precision. Of the several instruments, only two widely used ones are described here: the **180° magnetic deflection type mass spectrometer**, as exemplified by the MS10 mass spectrometer of Associated Electrical Industries (Manchester) Ltd.; and the **omegatron**. Of the other instruments, not to be described, probably the most

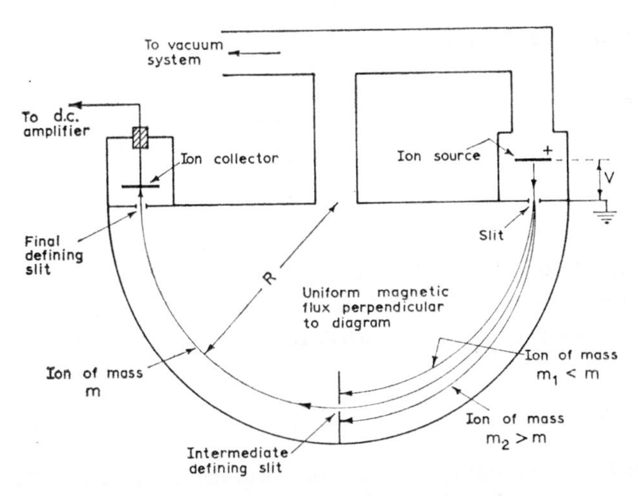

Fig. 65. Principle of 180° magnetic deflection mass spectrometer.

promising is the **quadrupole mass filter** (Günther [88]); commercial versions are made by Varian Associates Ltd., and by Atlas-Werke A.G., Bremen.

In the magnetic deflection mass spectrometer, the gas is ionized usually by impact of electrons from a thermionic filament, which are accelerated to a positive electrode. This is done within a confined ion source, generally following the pattern originated by Nier [89]. The positive ions produced are extracted from an outlet slit of the ion source, accelerated by a p.d. of V between this slit and a second parallel slit, and then directed into a uniform magnetic field of flux density B, the field lines of which are perpendicular to the ion paths (Fig. 65). A positive ion of mass m and charge ne, where e is the electronic charge and n is an integer, will thereby undergo deflection in this field through a path along the arc of a circle of radius R given by

$$\frac{mv^2}{R} = Bnev \tag{2.16}$$

where v is the velocity of the ion leaving the second slit. As this velocity is acquired by accelerating the ion through the p.d. V,

$$Vne = \tfrac{1}{2}mv^2 \tag{2.17}$$

From equation (2.17),

$$v = \sqrt{\left(\frac{2Vne}{m}\right)}$$

Substituting in equation (2.16), therefore,

$$m = \frac{BneR}{\sqrt{(2Vne/m)}}$$

$$\therefore\ m = \frac{B^2R^2ne}{2V} \tag{2.18}$$

If a particular, fixed value of the radius R of the circular path in the magnetic field is defined by slits, and the positive ions which emerge from the final, defining slit are collected at an electrode, at which the current they produce is measured, for a constant value of the magnetic flux density B,

$$m = \frac{kne}{V} \tag{2.19}$$

where k is a constant. This equation will decide the mass, m, of a positive ion which produces the maximum output current. For singly-charged positive ions, which occur the most frequently, n is unity.

If, therefore, the ion-accelerating p.d. V is varied, a graph can be plotted of positive ion output current against V, or $1/V$, which will show a series of peaks; each peak corresponding to the presence of positive ions of a mass-to-charge ratio m/ne given by equation (2.19) (where k is a constant of the instrument). Considering, for simplicity, the case where n is unity, a **mass spectrum** will therefore be obtained with each peak corresponding to a particular mass. Further, the relative peak heights will depend on the relative abundance of the various ions present. As the mass of the singly-charged positive ion equals that of the corresponding atom or molecule, except for the very small correction due to the mass of the single electron, the atoms or molecules present are identified from the masses recorded.

In the 180° magnetic deflection mass spectrometer, the ion beam is deflected through an angle of 180° in the magnetic field, so the beam describes a semicircle. Other angles of deflection: 60°, 90°, and 120°, have been used in several designs, following the pioneer work of Dempster [90] and Nier, but the 180° instrument is the most popular in application to the analysis of residual gases in vacuum practice.

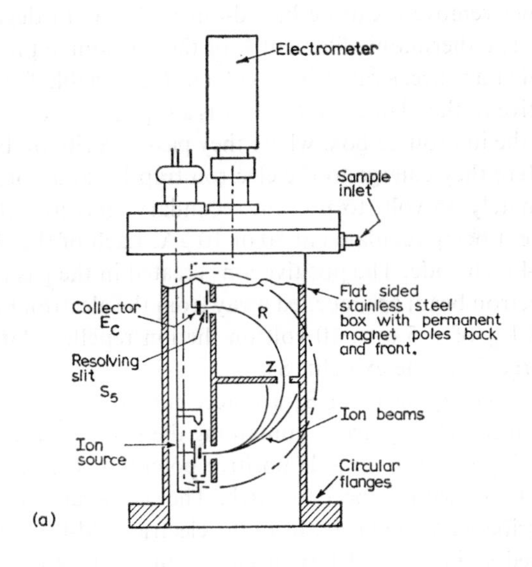

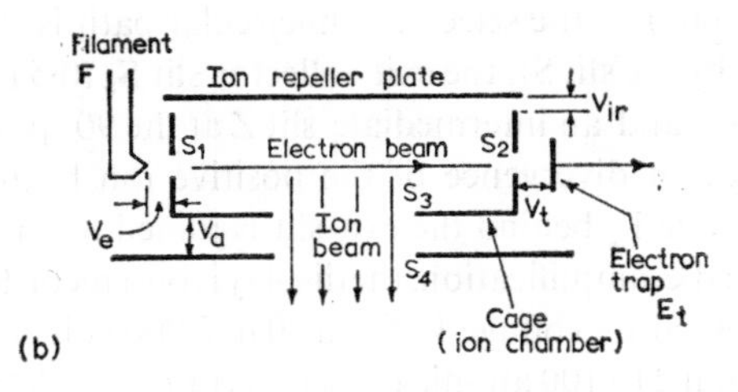

Fig. 66. (*a*) The Associated Electrical Industries Ltd., mass spectrometer MS10. (*b*) The ion source of the MS10.

Within the semicircular path, focusing of the ion beam takes place whereby the emergent ions at the collecting electrode form an approximate image of the slit, to which the ions from the ion source are initially accelerated.

Excellent practice in the design of a 180° instrument is typified by the Associated Electrical Industries, Ltd., MS10 mass spectrometer (Fig. 66*a*). The envelope is a stainless steel box with circular end flanges; the box is between the pole-pieces of a permanent magnet, which provides a magnetic flux density of 2,000 gauss (0·2 weber per sq metre) in a gap of width 1¼ inch. The ion source (Fig. 66*b*) electrodes and the ion collector system are mounted on cover plates, which are sealed to the flanges by gold-wire seals (section 3.8), so that ready demountability is assured and the instrument, with the magnet removed, can be baked-out at 400°C to degas it under vacuum. The thermionic filament F of the ion source produces electrons which are accelerated by a voltage V_e, variable from 5 to 105 volt relative to the filament F (70 volt is a typical value), to the slit S_1 and into the ion source box, where they pass in a ribbon beam to the slit S_2. Here they emerge to the electron trap E_t, at a potential V_t of approximately 35 volt, to prevent secondary emission, the electron trap current being regulated at 50 or 10 μA. Each of the slits, S_1 and S_2, is 0·04 inch wide. The positive ions created in the gas in the wake of the electron beam are repelled away from the electron beam by the potential V_{ir} of -5 to $+10$ volt on the ion repeller plate, so these ions emerge from the exit slit, S_3.

The ions emerging from S_3 are accelerated by the p.d. V_a to a parallel slit S_4. The electrode containing the slit S_4 is at earth potential, S_3 being set at a variable positive potential, V_a, of between 40 and 2,000 volt with respect to earth. The accelerated positive ions then describe semicircular paths in the electric-field-free space within the earthed stainless steel box, itself within the magnetic field. The radius R (Fig. 66*a*) of the selected semicircular path is 2 inch (5 cm), and is defined by the slit S_4, the exit collector slit S_5 (0·5 inch long and 0·02 inch wide), and an intermediate slit Z at the 90° position, which limits the angle of divergence of the positive ion beam to 4°. The collector electrode E_c behind the exit slit is joined to an electrometer amplifier; after d.c. amplification, the display is on a recorder. Variation of the ion-accelerating voltage V_a from 40 to 2,000 volt enables masses in the range from 2 to 100 atomic mass units (a.m.u., where 1 a.m.u. = $\frac{1}{12}$ of the mass of the carbon 12 atom) to be scanned. With an electron trap current of 50 μA, an electron trap potential V_t of 70 volt, and the ion repeller voltage at 1 volt, the sensitivity of the MS10 varies from a minimum of $0{\cdot}89 \times 10^{-5}$ amp per torr for helium to a maximum of $5{\cdot}6 \times 10^{-5}$ amp per torr for argon, with a value of $3{\cdot}6 \times 10^{-5}$ amp per torr for nitrogen; this last figure corresponds to a sensitivity of 0·7 per torr. The maximum operating pressure that can be used in the spectrometer is 10^{-4} torr. The mass resolving power, $m/\Delta m$ (where Δm corresponds to the mass difference that can be distinguished for a 1% valley between neighbouring peaks), is approximately 50. A 1% valley means here that two adjacent peaks of the same height would each contribute 0·5% to the valley midway between the peaks.

Typical 'cracking' patterns obtained with the MS10 for nitrogen, oxygen, carbon monoxide and methane, are shown in Fig. 67. The presence of the molecule, the separated atoms, and the isotopes are clearly shown.

This instrument has been extensively used for analysing the residual gases in vacuum systems under a wide variety of conditions, showing

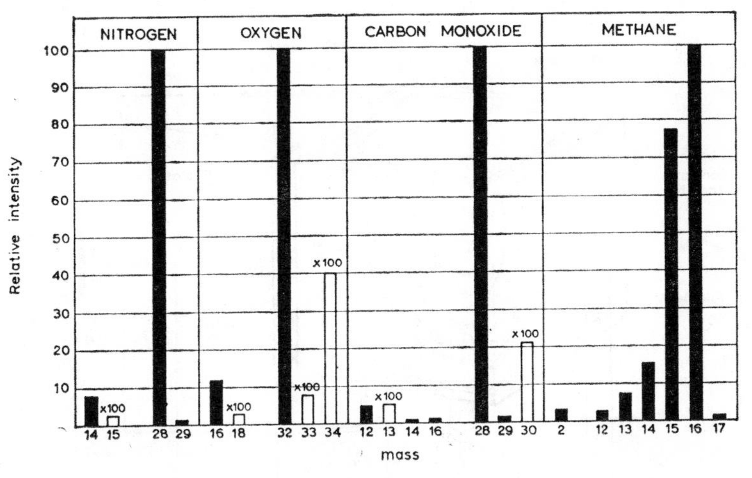

Fig. 67. Some typical cracking patterns obtained with the MS10.

the effects of different types of pump, bake-out, nature of envelope material, activation and operation of thermionic cathodes, etc., over the total pressure range from 10^{-4} to 10^{-10} torr, with the ability to detect partial pressures of 10^{-11} torr. By careful calibration or use of data provided by the manufacturer, partial pressures can be estimated from the peak heights in the characteristic cracking pattern of the gas concerned.

The **omegatron**, introduced originally by Sommer, Thomas, and Hipple [91] and first adapted for gas analysis *in vacuo* by Alpert and Buritz [92], is similar in principle to the cyclotron. Typically, a 2 cm cube box is formed from a platinum-iridium or a gold-plated, non-magnetic, copper-nickel alloy, or stainless steel sheet, with the top and bottom electrodes, P and P_1, electrically isolated from the box walls (Fig. 68). This structure is mounted in a glass envelope. Electrons from a thermionic filament F are accelerated through an aperture A_1, into the box, to form an axial beam which then leaves through an aperture A_2 in the opposite wall; then it is monitored on an electron collector electrode E_t. A uniform magnetic field of flux density B is maintained by a permanent magnet around the envelope, with the field lines parallel to the electron beam passing from A_1 to A_2. A radio-frequency p.d. of 0·5 to 2 volt r.m.s. is maintained across the electrodes P and P_1, and an ion collector electrode, E_i, is inserted within a slot through one of the plates, say P_1.

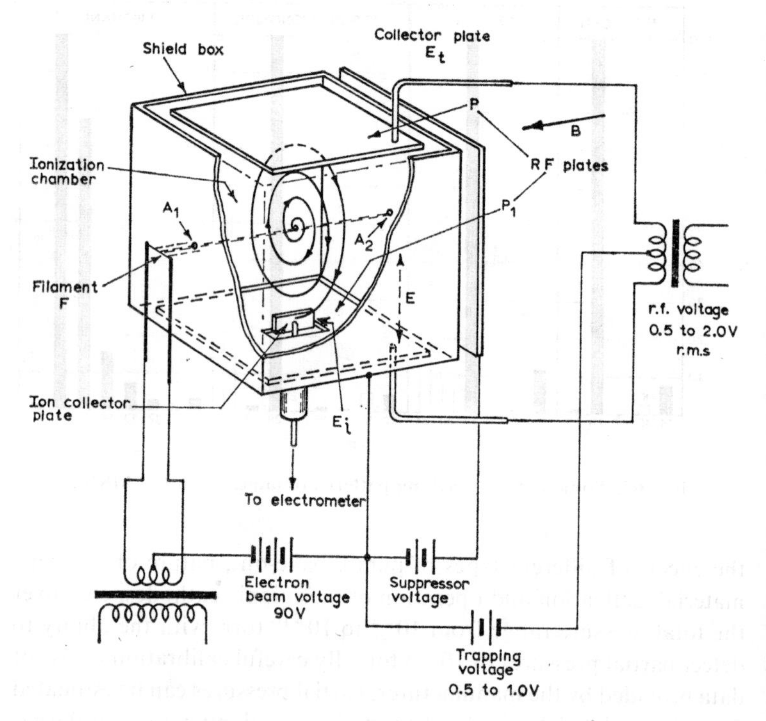

Fig. 68. The principle of the omegatron.

Positive ions are created in the residual gas in the region of the electron beam. During the half-cycle when P is negative with respect to P_1, these ions will move towards P. They may not reach P, however, because they are curved into a path of radius R by the action of the perpendicular magnetic field. R is given for a singly-charged ion by

$$R = mv/Be$$

where v is the mean velocity of the ions.

The time, $T/2$, taken by these ions to describe a semicircle is

$$T/2 = \pi R/v = \pi m/Be$$

which is independent of R and of v. If, therefore, the period of the alternating voltage across PP_1 is made equal to T, a particular ion of mass-to-charge ratio m/e will just finish describing a semicircular path as the p.d. across PP_1 is reversed. So the ions are now directed towards P_1, where, though their velocity is increased, they will now describe a semicircular path of larger radius but in the same time, $T/2$. The path of a positive ion is consequently a spiral terminating at the ion collector E_1, provided that

$$\frac{m}{e} = \frac{BT}{2\pi} = \frac{B}{\omega} \tag{2.20}$$

where ω is the pulsatance of the alternating p.d. across PP_1, and m/e the mass-to-charge ratio of the so-called resonant ion.

With a given value of B, say 2,500 gauss, variation of the frequency $f(=\omega/2\pi)$ from 50 kc per sec to 3 Mc per sec will enable singly-charged positive ions to become resonant at particular frequencies, so giving peak ion currents for values of m/e in the range from $2{,}500/(2\pi \times 5 \times 10^4)$ to $2{,}500/(2\pi \times 3 \times 10^6)$, i.e. from 8×10^{-3} to $1{\cdot}33 \times 10^{-4}$. As $e = 1{\cdot}6 \times 10^{-20}$ e.m.u. for a singly-charged ion, this corresponds to masses from $1{\cdot}28 \times 10^{-22}$ to $2{\cdot}23 \times 10^{-24}$ gram, which includes masses in atomic units (1 a.m.u. $= 1{\cdot}66 \times 10^{-24}$ gram) from 75 to 2, i.e. from selenium 75 to deuterium and molecular hydrogen at 2.

A simplified operating circuit for the omegatron (Fig. 68) shows typical values of the potentials used. Note the positive trapping potential of about 0·5 to 1·0 volt applied to the box, relative to one of the radio-frequency plates, to produce an electric field, which reduces the loss of ions in the direction of the magnetic field.

The omegatron can be baked at 450°C and the electrodes degassed by induction heating (p. 244). With an electron current of 4 μA a sensitivity for nitrogen of 10 per torr is obtainable, which is comparable with the hot-cathode ionization gauge and more than 10 times that of the magnetic deflection mass spectrometer. The resolving power decreases with the mass of the ion.

The small mass spectrometer and the omegatron have similar

functions in providing an analysis of gases at total pressures in the range from 10^{-4} to 10^{-10} torr. The resolution of the mass spectrometer is superior, and it can be more readily calibrated to read actual partial pressures of the constituent gases. Furthermore, the interpretation of the cracking patterns is easier, as fewer problems of interpreting the significance of mass-to-charge ratio appearances arise. The chief advantage of the omegatron is its higher sensitivity.

2.10. *The Ranges of Vacuum Gauges*

The range of pressures over which the various vacuum gauges described will operate usefully are denoted by the chart of Fig. 69. Methods of calibrating vacuum gauges are described in section 3.9.

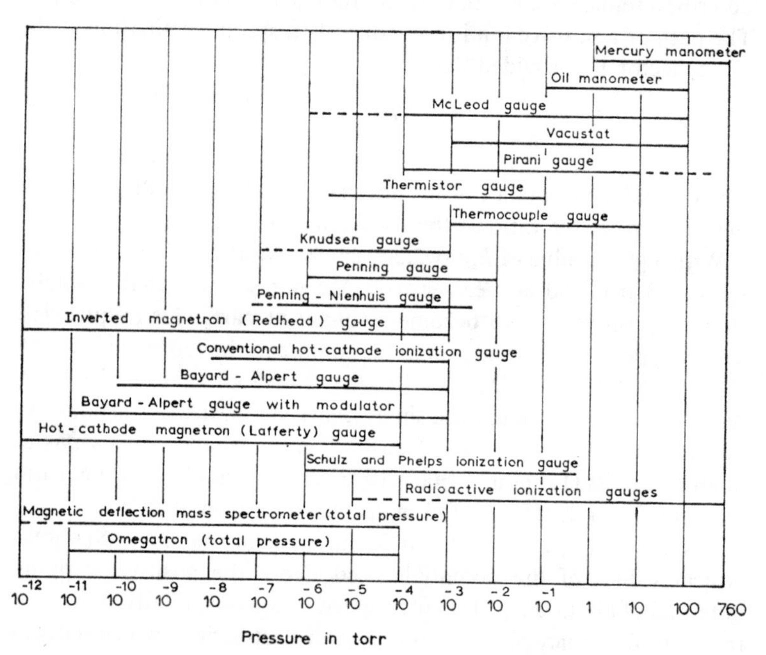

Fig. 69. The ranges of vacuum gauges.

CHAPTER THREE

VACUUM SYSTEMS, COMPONENTS, AND LEAK DETECTION

3.1. *Viscous and Molecular Flow of Gases*

Before describing the practical methods of setting up vacuum systems, the laws governing the flow of gas through apertures and tubes at low pressures must be studied.

Suppose two parallel solid surfaces exist at a distance d apart, forming boundaries in a gas at a pressure p, such that the m.f.p. is L. Three possible cases arise: (*a*) $L<d$; (*b*) $L=d$ approximately; (*c*) $L>\mathrm{d}$.

In case (*a*), collisions between the gas molecules themselves will be more important than collisions of the gas molecules with the boundaries. The gas flow between the boundaries is then said to be **viscous.** Viscous flow will prevail at pressures depending on the order of size of *d*. For example, if $d=5$ cm, viscous flow will predominate for values of L less than, say, 1 cm. For nitrogen, the pressures concerned will therefore be above 5×10^{-3} torr (equation 1.4).

In the frequently encountered practical case of a long uniform cylindrical tube of radius r and length l, where l is considerably greater than r, Poiseiulle's equation applies to viscous flow, i.e.

$$p_1V_1 = p_2V_2 = \frac{p_1^2-p_2^2}{16\eta l}\pi r^4 \tag{3.1}$$

where p_1 is the gas pressure at the inlet to the tube, p_2 the pressure at the outlet, V_1 the volume of gas entering the tube per sec at pressure p_1, V_2 the volume emerging from the outlet at pressure p_2, and η is the coefficient of viscosity of the gas.

From Boyle's law, as the mass of gas passing any cross-section of the tube per sec is constant, $p_1V_1=p_2V_2$. It is valuable to call pV the **throughput,** Q.

The **impedance** of the tube to flow of gas, Z, is defined by

$$Z = \frac{p_1-p_2}{Q} = \frac{1}{U} \tag{3.2}$$

where the reciprocal of the impedance is called the **conductance** U of the tube. From equation (3.1), it is seen that

$$U = \frac{Q}{p_1 - p_2} = \left(\frac{p_1 + p_2}{16\eta l}\right)\pi r^4$$

It is convenient to put $(p_1 + p_2)/2 = \bar{p}$, the average pressure in the tube, then,

$$U = \frac{\pi r^4 \bar{p}}{8\eta l}$$

$$= \frac{\pi d^4 \bar{p}}{128\eta l} \tag{3.3}$$

where $d = 2r$, the tube diameter. The rate of gas flow in quantity is given by $Q = U\Delta p$, where Δp is the pressure difference.

The coefficient of viscosity η of a gas is usually expressed in the c.g.s. unit – the poise. For air at 20°C, the viscosity of air is 180 micropoise.

As η is given in poise, p must be in dyne per sq cm (microbar), with d and l in cm. 1 torr equals 1,333 microbar. Equation (3.3) therefore becomes

$$U = \frac{\pi d^4 \bar{p} \times 1{,}333}{128\eta l}$$

where $\bar{p}$ is in torr and η in poise.

$$\therefore\ U = \frac{10{\cdot}4\pi d^4 \bar{p}}{\eta l} \tag{3.4}$$

For air at 20°C, substituting $\eta = 180 \times 10^{-6}$ poise,

$$U = \frac{5{\cdot}8\pi d^4 \bar{p}}{l} \times 10^4 \text{ cm}^3 \text{ sec}^{-1}$$

$$= \frac{182\bar{p}d^4}{l} \text{ litre sec}^{-1}$$

For example, if a cylindrical tube has a length of 50 cm and a diameter of 0·5 cm, with an air pressure at its inlet of 0·5 torr and at its outlet of 0·1 torr, $\bar{p}$ is 0·3 torr and

$$U = \frac{182 \times 0{\cdot}3 \times 0{\cdot}25}{50} = 0{\cdot}27 \text{ litre sec}^{-1}$$

As $Q = (p_1 - p_2)U$, in this example

$$Q = (0{\cdot}5 - 0{\cdot}1)0{\cdot}27$$

$$= 0{\cdot}108 \text{ torr litre sec}^{-1}$$

Note the unit of Q, *torr litre per sec*, i.e. a quantity of gas specified in litre at a pressure of 1 torr flowing in 1 sec.

Further, as $Q = pV$, the volume of gas emerging at the outlet per sec is

$$V_2 = \frac{Q}{p_2} = \frac{0{\cdot}108}{0{\cdot}1} = 1{\cdot}08 \text{ litre sec}^{-1}$$

at a pressure of 0·1 torr.

Case (*b*), where $L = d$ approximately, is an awkward one where intermolecular and wall collisions both play an important part. In this intermediate case, known as *Knudsen flow*, semi-empirical equations have been derived but will not be considered here.

Case (*c*) is a most important one in high vacuum technique. Now the collisions of the molecules with the walls are the chief factor deciding the gas flow. The gas flow is then said to be **molecular**. For example, if $d = 2$ cm, molecular flow would predominate for m.f.p.'s in excess of, say, 5 cm, corresponding to nitrogen pressures of 10^{-3} torr and below. Such low pressures are known as **molecular pressures,** taken as less than 10^{-3} torr in general, though for boundaries in the gas separated by a considerable distance, this assumption may be unjustified and intermediate flow concerned.

3.2. *Molecular Effusion Through an Aperture*

A first important case is that of **molecular effusion of gas through an aperture in a thin wall** connecting two containers of dimensions large compared with that of the aperture. Let p_1 be the pressure in one of these chambers and p_2 that in the other, where $p_1 > p_2$ and both are in the molecular region.

From Knudsen's work in the kinetic theory of gases, it is shown that the number of molecules N impinging on unit area of a boundary in the gas per sec is given by

$$N = \tfrac{1}{4} n\bar{v}$$

where n is the number of molecules per unit volume, i.e. the molecular density.

Put

$$\bar{v} = \sqrt{\left(\frac{8\mathrm{R}T}{\pi M}\right)}$$

a result from Maxwell's distribution law for the velocities of molecules in a gas of molecular weight M and at absolute temperature T, R being the gas constant per mole, and

$$p = nkT = n\mathrm{R}T/\mathrm{N}$$

(*See* equation 1.1, where $k = R/N$, N being Avogadro's number.)

$$\therefore\ N = \frac{1}{4}\frac{pN}{RT}\sqrt{\left(\frac{8RT}{\pi M}\right)} = \frac{pN}{\sqrt{(2\pi MRT)}} \qquad (3.5)$$

If the aperture has an area A, the number of molecules N_1A impinging on it from the higher pressure side is therefore $p_1NA/\sqrt{(2\pi MRT)}$, whilst the number N_2A impinging from the lower pressure side is $p_2NA/\sqrt{(2\pi MRT)}$. The net flow of molecules per sec, i.e. the molecular effusion through the aperture, is therefore given by

$$(N_1 - N_2)A = (p_1 - p_2)NA/\sqrt{(2\pi MRT)}$$

As N molecules occupy a specific volume V at pressure p, where $pV = RT$, $(N_1 - N_2)A$ molecules will correspond to a throughput of

$$\left[\frac{(N_1 - N_2)A}{N}\right]RT$$

$$\therefore\ Q = \frac{RT}{N}\left[\frac{(p_1 - p_2)NA}{\sqrt{(2\pi MRT)}}\right]$$

$$= \frac{(p_1 - p_2)A}{\sqrt{(2\pi M/RT)}} \qquad (3.6)$$

From equations (3.2) and (3.6), it follows that the molecular conductance U of the aperture is given by

$$U = \frac{Q}{(p_1 - p_2)} = \frac{A}{\sqrt{(2\pi M/RT)}}$$

Substituting $R = 8{\cdot}314 \times 10^7$ erg per degC per mole, and putting $T = 293°K$ (20°C) as an average room temperature,

$$U = A/\sqrt{(2\pi M/8{\cdot}314 \times 10^7 \times 293)}$$

$$= 6{\cdot}25 \times 10^4 A/\sqrt{M}\ \text{cm}^3\ \text{sec}^{-1}$$

$$= 62{\cdot}5A/\sqrt{M}\ \text{litre sec}^{-1} \qquad (3.7)$$

In the case of air, considered as a nitrogen-oxygen mixture of molecular weight $M = 29$, equation (3.7) becomes

$$U = 62{\cdot}5A/\sqrt{29} = 11{\cdot}6A\ \text{litre sec}^{-1}$$

Air at molecular pressures (i.e. below 10^{-3} torr) cannot therefore traverse an aperture in a thin wall at a rate exceeding 11·6 litre per sec.

3.3. *Molecular Flow through a Cylindrical Tube*

The molecular flow of gas through a long cylindrical tube, in the case where the length l exceeds considerably the diameter d, and d is significantly less than the m.f.p. of the gas in the tube, can be calculated by considering an element of length Δl of this tube. This will have a cylindrical surface area of $\pi d.\Delta l$, and $\frac{1}{4}n\bar{v}$ molecules will impinge on unit area of this surface per sec. The momentum ΔG transferred per sec to this element of surface will therefore be given by

$$\Delta G = \pi d.\Delta l.\tfrac{1}{4}n\bar{v}.mu.f \qquad (3.8)$$

where m is the mass of a molecule of the gas having a mean velocity of drift through the tube of u, and hence momentum mu. The fraction f has a value depending on the gas and the nature of the walls; for want of an experimental value, f is usually taken to be unity.

Let p be the pressure inside this element of the tube of length Δl, Δp, the pressure difference across Δl, and so the force acting across the element, will be $\Delta p.\pi d^2/4$. Equating this force to the momentum per sec imparted to the cylindrical walls,

$$\frac{\Delta p.\pi d^2}{4} = \frac{\pi d.\Delta l.n\bar{v}\,mu}{4}$$

$$\therefore\ u = \frac{\Delta p.d}{mn\bar{v}.\Delta l}$$

The mass Q_m of gas flowing through the tube element per sec is hence given by

$$Q_m = \frac{\pi d^2}{4}u\rho = \frac{\pi d^3.\Delta p.\rho}{4mn\bar{v}.\Delta l}$$

where ρ is the density of the gas at pressure p.

But

$$\rho = mn,$$

$$\therefore\ Q_m = \frac{\pi d^3.\Delta p}{4\bar{v}.\Delta l}$$

In the case of a tube of finite length l, where p_1 is the pressure at the inlet, p_2 the pressure at the outlet, and $p_1 > p_2$, the mass of gas passing through any cross-section of the tube per sec must be constant, because there is no accumulation of gas in a continuous flow. Therefore, the mass throughput

$$Q_m = \frac{\pi d^3(p_1 - p_2)}{4\bar{v}l}$$

This mass of gas Q_m per sec will correspond to a volume V of gas equal to Q_m/ρ, where ρ is the density at a pressure p. Therefore, the volumetric throughput Q is given by

$$Q = \frac{Q_m p}{\rho} = \frac{\pi d^3(p_1 - p_2)p}{4\bar{v}l\rho} \qquad (3.9)$$

From Boyle's law, p/ρ is a constant, and applying equation (1.1) it is given by

$$p/\rho = nkT/nm = kT/m = RT/M$$

where M is the molecular weight of the gas.

Further, from equation (1.7)

$$\bar{v} = \sqrt{\left(\frac{8RT}{\pi M}\right)}$$

Substituting these expressions for p/ρ and $\bar{v}$ in equation (3.9) gives

$$Q = \frac{\pi d^3(p_1 - p_2)RT}{4\sqrt{\left(\frac{8RT}{\pi M}\right)}Ml} = \frac{\pi d^3(p_1 - p_2)}{16l}\sqrt{\left(\frac{2\pi RT}{M}\right)}$$

$$= \frac{1}{5{\cdot}1}\sqrt{\left(\frac{2\pi RT}{M}\right)}\frac{d^3}{l}(p_1 - p_2) \qquad (3.10)$$

This result, obtained by a straightforward method, is not quite correct. A more elaborate theory takes into account, in detail, the fact that when a gas molecule strikes the wall it will leave it in any direction, where any one direction for a single molecule has the same probability as any other direction. The full theory gives a numerical factor of 1/6 in place of 1/5·1 in equation (3.10), so the accepted equation is

$$Q = \frac{1}{6}\sqrt{\left(\frac{2\pi RT}{M}\right)}\frac{d^3}{l}(p_1 - p_2) \qquad (3.11)$$

By equation (3.2) therefore, the molecular conductance of the tube U, is given by

$$U = \frac{1}{6}\sqrt{\left(\frac{2\pi RT}{M}\right)}\frac{d^3}{l} \qquad (3.12)$$

It is of immediate importance in practice, to note that the conductance U increases with the cube of the tube diameter and decreases with increase of length. To ensure that a tube is able to offer little opposition to gas flow and so not restrict pumping speed, *it is essential to employ large-bore short tubes*, and where a large-bore diameter is more important than shortness.

Inserting $R=8{\cdot}314\times10^7$ erg per degC per mole, and $T=293°K$ (i.e. 20°C) in equation (3.12) gives

$$U=\frac{1}{6}\sqrt{\left(\frac{2\pi\times8{\cdot}314\times10^7\times293}{M}\right)}\frac{d^3}{l}$$

$$=\frac{6{\cdot}5\times10^4}{\sqrt{M}}\frac{d^3}{l}$$

where U is in cu cm per sec, provided that d and l are in cm. It is usual to express U in litre per sec, where

$$U=\frac{65}{\sqrt{M}}\frac{d^3}{l}\text{ litre sec}^{-1} \tag{3.13}$$

Note that the conductance is inversely proportional to the square root of the molecular weight of the gas, and so will be larger for hydrogen ($M=2$) than for nitrogen ($M=28$) by a factor of $\sqrt{14}$.

For air, put $M=29$

$$\therefore\ U=\frac{65}{\sqrt{29}}\frac{d^3}{l}=\frac{12{\cdot}1d^3}{l}\text{ litre sec}^{-1} \tag{3.14}$$

It is convenient here to put $d=2r$, where r is the radius of the tube in cm, then,

$$U=\frac{12{\cdot}1(2r)^3}{l}=\frac{100r^3}{l}\text{ litre sec}^{-1}\text{ (approximately)}$$

which is easily remembered.

For a tube of length l which is not long compared with its diameter d, it is essential to take into account the fact that the inlet aperture to the tube has an impedance to gas flow which is significant compared with that of the tube itself. In the previous case, where $l\gg d$, this aperture has been neglected. When the tube inlet is from a vessel of comparatively large dimensions, the molecules of gas in the vessel, with their random kinetic motion, will have to 'find' the tube aperture to get into the tube. A simple way of taking this into account which gives a result accurate enough for most practical purposes, though the procedure can be criticised on theoretical grounds, is to consider the overall tube conductance to be given by the sum of an aperture conductance, as given by equation (3.7), in series with a tube conductance, as given by equation (3.13). For conductances in series (compare with electrical case), the equation is

$$\frac{1}{U}=\frac{1}{U_{ap}}+\frac{1}{U_{tube}}$$

where U is the overall conductance of the tube, U_{ap} the conductance of its inlet aperture, and U_{tube} that of the tube itself. From equations (3.7) and (3.13) for a gas of molecular weight M at 20°C,

$$\frac{U_{ap}}{U_{tube}} = \frac{62 \cdot 5\pi d^2/4}{65d^3/l} = \frac{3l}{4d}$$

$$\therefore\ U = \frac{U_{tube} \times U_{ap}}{U_{tube} + U_{ap}}$$

$$= \frac{\left(\frac{65}{\sqrt{M}}\frac{d^3}{l}\right)^2 \frac{3l}{4d}}{\frac{65}{\sqrt{M}}\frac{d^3}{l}\left(1+\frac{3l}{4d}\right)}$$

$$= \frac{65d^3}{\sqrt{M}}\left(\frac{1}{l+4d/3}\right) \text{ litre sec}^{-1} \qquad (3.15)$$

where l and d are in cm.

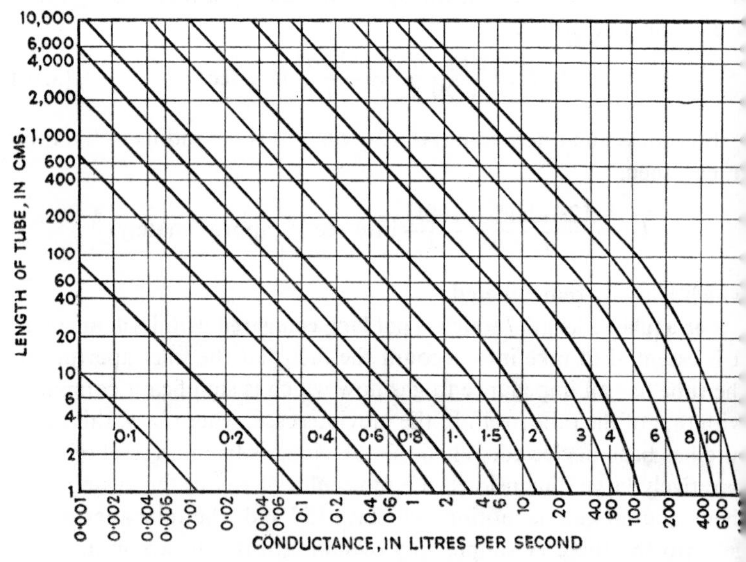

Fig. 70. Conductances of uniform cylindrical tubes for air at 20°C. (Parameter is the diameter of the tube in cm.)

For air, putting $M = 29$,

$$\therefore\ U = \frac{12 \cdot 1d^3}{l+(4d/3)} \qquad (3.16)$$

This is a simple equation to use to calculate molecular conductances of uniform cylindrical tubes; however, for convenience, the curves of Fig. 70, based on equation (3.16), enable the conductance of a variety of tubes of various lengths and diameters to be obtained quickly.

3.4. *Effect of Connecting Tube on Pump Speed*

The most important example where molecular conductance has to be calculated is when a tube is used to connect a chamber to the intake port of a vapour pump. The speed of a pump has already been defined (p. 6) and is usually expressed for vapour pumps in litre per sec. If S is this pump speed and U the conductance of a tube which is connected in series with it,

$$\frac{1}{S_E} = \frac{1}{S}+\frac{1}{U}$$

where S_E is the effective pump speed at the chamber

$$\therefore \; S_E = \frac{SU}{S+U} \tag{3.17}$$

It is immediately apparent that S_E cannot exceed S and, further, cannot exceed U. If U is small because a long narrow tube is used, practically no increase of S_E is obtained by attempting to make S very large. In other words, if U is inevitably small (e.g. in pumping an electronic tube through a narrow glass tube) there is no point in choosing other than a small pump; a large pump would be wasteful.

A vapour diffusion pump with an intake port diameter of 5 cm will have a pumping speed of about 60 litre per sec. If it is connected to a chamber by a tube of diameter 5 cm and length 20 cm, the conductance of this tube for air at 20°C is given by equation (3.16) (or by reference to Fig. 70) to be

$$U = \frac{12{\cdot}1\times 5^3}{20+(4\times 5)/3} = 56{\cdot}5 \text{ litre sec}^{-1}$$

This is the greatest speed that can be obtained through this tube, no matter how large the pump.

The effective speed, S_E, at the chamber, with the 60 litre per sec pump, is given by equation (3.17) as

$$S_E = \frac{60\times 56{\cdot}5}{60+56{\cdot}5} = 29 \text{ litre sec}^{-1}$$

If a 2 cm tube of the same length had been used instead of a 5 cm one,

$$U = \frac{12{\cdot}1\times 2^3}{20+(4\times 2)/3} = 4{\cdot}25 \text{ litre sec}^{-1}$$

so the maximum possible effective speed, even if a pump of infinite speed were available, would be less than $\frac{1}{6}$ that with the 5 cm tube.

The great importance of tube diameter in high vacuum work is thus apparent. Indeed, the use of long narrow tubes should be avoided as far as is practicable.

3.5. *The Use of Oil-sealed Mechanical Pumps*

The production of low pressures in the range from 1 atmosphere down to 10^{-2} torr is most frequently undertaken with an oil-sealed mechanical pump alone. Such pumps will operate reliably for many years, but a number of important aspects of practice need to be observed to ensure satisfactory results.

When using oil-sealed mechanical pumps either alone or in the backing stage to a vapour pump, care must be taken to ensure that, if the pump is left switched-off, oil is not sucked back from the pump into any evacuated vessel connected to its intake port. To prevent this, some manufacturers fit the pump with a self-sealing oil valve in the intake, but older pumps often do not have this provision, and, in any case, such valves are sometimes unreliable over long periods. The best precaution is to isolate the evacuated system from the pump by closing a greased, glass stopcock (section 3.6), or a vacuum valve (section 3.7), when the plant is shut-down, switch-off the pump and admit air to it by a second, small, glass stopcock, or by a metal, air-inlet valve. A very good way of arranging these provisions, with automatic operation on switching-off the pump motor, is by the use of a magnetic valve.

The removal of water vapour from a vacuum system is of great importance: the mechanical pump will not cope with it satisfactorily unless a gas-ballast type (section 1.4) is used. A drying agent placed in a container attached to the mechanical pump intake port will improve the performance. This is especially useful for pumps without gas-ballast, and in laboratory systems (as opposed to larger scale industrial ones) is not to be despised even if gas-ballast is provided.

The most efficient of the various drying agents which are convenient is phosphorus pentoxide (freedom from arsenic is essential). In a glass system, this is conveniently placed in a glass trough, inserted in a glass tube furnished with a greased cone-joint stopper which, when removed, permits replacement of the pentoxide (Fig. 71*a*). A convenient phosphorus pentoxide trap for a metal system is shown in Fig. 71(*b*).

Phosphorus pentoxide hydrates to become phosphoric acid with a saturated vapour pressure of about 5×10^{-4} torr at room temperature. It is therefore not good practice to use this material as a drying agent in a high vacuum chamber evacuated by vapour pumps.

In using oil-sealed rotary pumps, especially of the larger sizes, certain provisions are desirable. Oil is entrained in the gas discharged to the atmosphere, particularly at higher intake pressures to the pump, so an oil separator, oil mist filter, or catch-pot in the discharge outlet, with the exhaust gas vented to outside the building, is needed

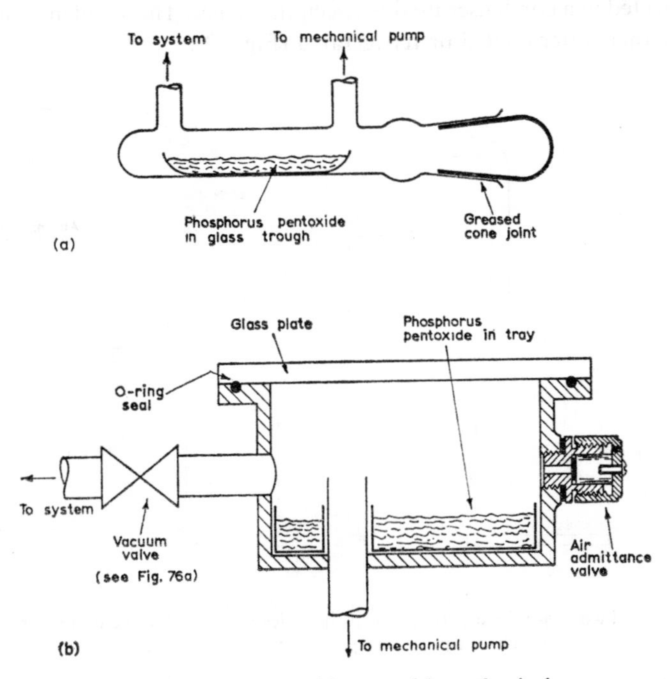

Fig. 71. Phosphorus pentoxide trap with mechanical pump.

(Fig. 72). If considerable quantities of water vapour are to be pumped at partial pressures exceeding 30 torr, a water-cooled condenser is advisable between the chamber and the intake port to the gas-ballast pump. The smaller sizes of gas-ballast two-stage rotary pumps usually have the two stages of the same capacity and only the second stage is gas-ballasted. As the compression in the first stage is then small, the chance of condensation, between the stages, of any water vapour present is small. With larger size pumps, however, considerable economy in size and cost results if two single-stage pumps are used in series, with the speed of the first-stage pump about 10 times that of the second pump, which discharges to the atmosphere. For example, if a speed of 1,000 litre per min is required at 1 torr, and the first stage produces a compression ratio of 10, the intermediate pressure between the stages is 10 torr, so the second-stage pump can be much smaller as it only has to pump at 100 litre per min at an intake pressure of 10 torr. Now, however, if water vapour is present at the intake, condensation between the stages is likely. The liquid resulting is collected in a condenser used between the stages. This condenser may be either water cooled or refrigerated (Fig. 72).

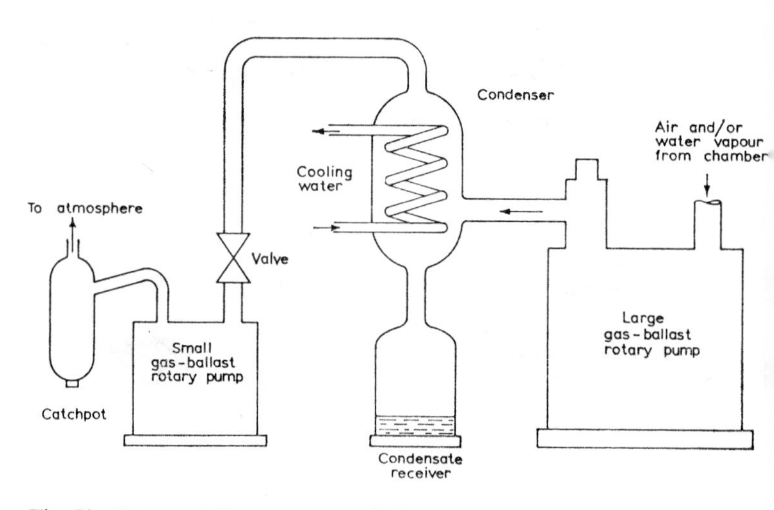

Fig. 72. Two gas-ballast rotary pumps in series with inter-stage condenser and catch-pot in discharge outlet to atmosphere.

The **pump-down time** required to reduce the pressure from p_1 to p_2 in a chamber of volume V by means of a pump of speed S may be calculated. Initially, the amount of gas in the chamber is of volume V at pressure p_1. After connecting the chamber with the pump, an amount of gas of volume $\mathrm{d}V$ will enter the pump in the short time $\mathrm{d}t$, where the pumping speed S is given by

$$S = \frac{\mathrm{d}V}{\mathrm{d}t}$$

From Boyle's law, the product pV must be constant at constant temperature during the increase of the volume by $\mathrm{d}V$. Therefore,

$$\frac{\mathrm{d}(pV)}{\mathrm{d}t} = 0$$

$$\therefore\ p\frac{\mathrm{d}V}{\mathrm{d}t}+V\frac{\mathrm{d}p}{\mathrm{d}t} = 0$$

$$\therefore\ S = \frac{\mathrm{d}V}{\mathrm{d}t} = \frac{-V}{p}\frac{\mathrm{d}p}{\mathrm{d}t}$$

or

$$S\,\mathrm{d}t = -V\frac{\mathrm{d}p}{p}$$

If p is p_1 at time t_1 and p_2 at a later time t_2, where $p_1 > p_2$, integration gives

$$S(t_2-t_1) = V\log_e\frac{p_1}{p_2} \qquad (3.18)$$

$$t_2-t_1 = \frac{V}{S}\log_e\frac{p_1}{p_2}$$

The pump-down time T, from $t_1=0$ to $t_2=T$, will hence be

$$T = \frac{2{\cdot}3V}{S}\log_{10}\frac{p_1}{p_2} \qquad (3.19)$$

If V is in litre and S in litre per min, the time T is given in min provided p_1 and p_2 are in the same units.

In using equation (3.19), the speed S must be the speed at the chamber and so take into account the effect of any restricting tubing between the chamber and the pump. Further, it is simple practice to assume S is constant. This is justifiable for an oil-sealed rotary pump over the pressure range from atmospheric down to about 10^{-1} torr, for a single-stage model, and 10^{-2} torr, for a two-stage one, assuming that the gas remains the same. Thus, a pump of effective speed at the chamber of 50 litre per min will pump down a volume of 10 litre from 760 torr to 10^{-1} torr in a time

$$T = \frac{2{\cdot}3\times 10}{50}\log_{10}\frac{760}{10^{-1}}\ \text{min}$$

$$= 1{\cdot}8\ \text{min}$$

The time, calculated by means of equation (3.19), to attain the lower pressures below 10^{-1} torr is usually too small, because the constitution of the air will change as the pressure is reduced and, in particular, the effect of water vapour begins to predominate in practice.

Generally, it is best to choose, if possible, an oil-sealed rotary pump considerably larger in speed than that estimated by equation (3.19), if quick exhaust is required to pressures below ten times the ultimate pressure of the pump.

VACUUM SYSTEMS

The numerous applications of vacuum technology, involving a wide range of pressures, pumping speeds, gases, processes, and particularly plant sizes, have led to the development of a multitude of vacuum systems and a variety of vacuum valves, pipe-line fittings, vacuum unions, lead-ins, and so on. As this range is too great to cover in a concise text, certain typical systems are described, selected for their usefulness and because they each indicate fundamental aspects of design. The components used are described in conjunction with each system. The emphasis is on small-scale laboratory plant; industrial plant is considered to be outside the present scope.

3.6. *Glass System with Mercury Vapour Pump*

A useful system for evacuating electron tubes and similar chambers down to pressures of 10^{-6} to 10^{-7} torr (Fig. 73) employs a glass mercury vapour diffusion pump. This pump is backed by a small two-stage mechanical gas-ballast rotary pump (free air displacement, say 30 litre per min) provided with a two-way stopcock S_1 and an air-inlet stopcock S_2. The intake port to the mercury vapour pump is connected to the glass chamber to be evacuated via a greased, glass stopcock, S_3, and a cold trap refrigerated with liquid air or liquid nitrogen. If a McLeod gauge is used to record the pressure, it may be connected between the stopcock S_3 and the cold trap. The vapour pressure of its mercury is then not exerted in the glass chamber or electron tube.

A **by-pass** or **roughing line** is convenient leading from one side of the two-way stopcock S_1 directly to a junction in the tubing between stopcock S_3 and the cold trap. This enables the glass chamber to be pumped to a backing pressure of about 10^{-2} torr, whilst the pre-evacuated mercury vapour pump is isolated and kept running. The saving in time in operation is considerable. A reservoir, of volume 3 to 5 litre, between the rotary backing pump and the discharge outlet of the vapour pump is worthwhile. This provides a reservoir volume maintained at backing pressure against which the mercury vapour pump operates when isolated.

The sequence in operation is briefly as follows, assuming the system is initially at atmospheric pressure. Close the stopcock S_1 and the air-inlet stopcock S_2. Switch on the pump and after 10 sec or so open S_1 to the reservoir and the mercury vapour pump. The initially closed stopcock, S_3, is then opened and the glass chamber evacuated to a

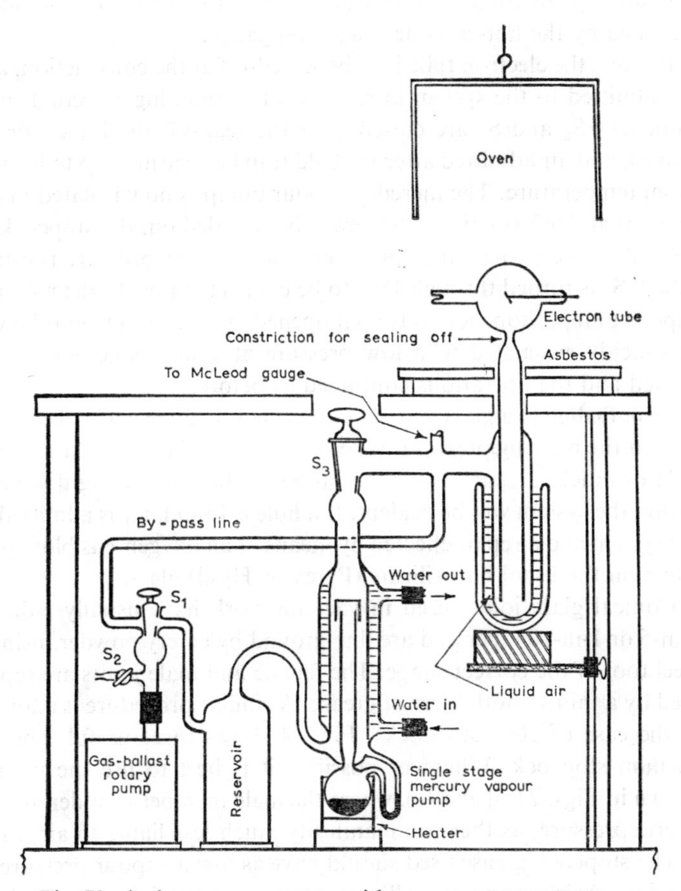

Fig. 73. A glass vacuum system with mercury vapour pump.

backing pressure as indicated by the McLeod gauge or by a satisfactory gaseous discharge set up by a Tesla coil. The liquid-air-filled Dewar flask has *not* yet been raised to surround the cold trap. The cooling water to the mercury vapour pump is turned on, and then the electric heater is turned on for its boiler. After about 20 min, the pressure will decrease to 10^{-5} to 10^{-6} torr, as recorded by the

McLeod gauge. The electron tube is then baked in the lowered bake-out oven to 450°C for borosilicate glass, or to 350°C for soda glass. Bake-out is usually for about 30 min, for a glass electron tube of volume not exceeding 2 litre. After the first 10 min of bake-out, the Dewar is raised so that the cold trap is refrigerated. After bake-out and subsequent cooling, a pressure of 10^{-6} torr or below should be recorded by the hot-cathode ionization gauge.

If, now, the electron tube is to be sealed-off at the constriction, and air admitted to the system in readiness for pumping a second tube, stopcocks S_3 and S_1 are closed after the seal-off, the liquid air removed, and air admitted after the cold trap has warmed up to be near room temperature. The mercury vapour pump is now isolated under vacuum and left running. The new tube is sealed on, the stopcock S_1 opened to the by-pass line and, when the backing pressure is established, S_1 is turned through 180° to be connected now to the mercury vapour pump. Stopcock S_3 is then opened, and the electron tube will be quickly evacuated to a low pressure at which bake-out can be started and the operations continued as before.

Leak finding in a glass system is conveniently done with a Tesla coil used at the backing pressure. When the Tesla coil probe is near a pin-hole or crack, a spark from the probe to the conducting discharge within the system will be evident. If a hole is found, air is admitted to the system and a repair effected by means of an oxygen-gas blow-pipe flame on the usual borosilicate (Pyrex or Hysil) glass.

Conical glass joints used in vacuum work have usually either a 1-in-5 or 1-in-10 taper and are fine-ground by emery powder, using a steel tool of the correct shape. The female and male joints are separated by a thin smooth layer of grease. A similar procedure is adopted in the case of glass stopcocks. Fig. 74 shows three useful types of vacuum stopcock. Wherever possible, it is best to use the models shown in Figs. 74(*b*) and (*c*), where the male member is under atmospheric pressure, as they are manifestly much less liable to air leaks.

The stopcock grease used should have as low a vapour pressure as possible. **Apiezon** and also **silicone greases** are specially prepared for this purpose. Apiezon L grease is supplied in metal tube containers. Its vapour pressure is 10^{-10} to 10^{-11} torr at 20°C, but this cannot be relied upon unless in an extremely clean condition. Its maximum service temperature is 30°C. In use, a thin film of the grease is worked between well-fitting cone or flange joints until no 'air-lines' remain. Apiezon M and N are similar, rather more viscous greases, the former having a vapour pressure of 10^{-7} to 10^{-8} torr at 20°C, and the latter 10^{-8} to 10^{-9} torr. A newer grease is Apiezon T, with a vapour pressure of 10^{-8} torr, and with the advantage that its maximum service temperature is 110°C. These greases must be kept free from contact with air when not in use; they give more trouble due to occluded air than due to high vapour pressure. A stopcock will give continual use for three months if properly greased. Silicone greases are remarkably stable from −40 to 200°C, with constant viscosity characteristics. The Dow Corning high vacuum grease, suitable for operating at pressures below 10^{-6} torr, is rather more viscous than the Apiezon greases at 20°C but functions very well, especially if the stopcock is liable to temperature extremes during use.

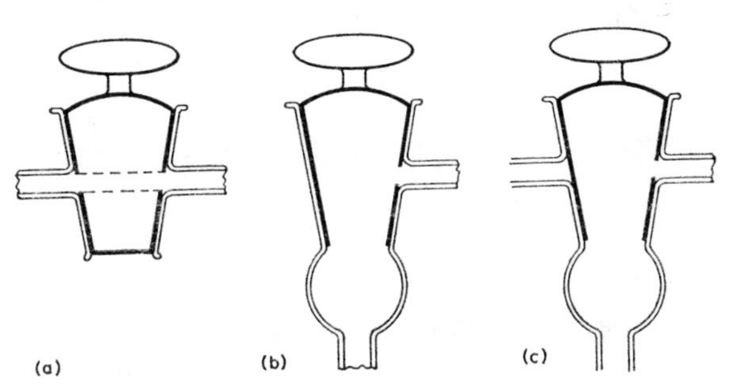

Fig. 74. Glass stopcocks.

3.7. *Demountable Metal System with Oil Vapour Pump*

A typical laboratory system for general purposes is one in which a bell-jar or similar vessel can be evacuated to about 10^{-5} torr by means of an oil vapour pump (Fig. 75). A pumping speed of at least 5 litre per sec for every litre capacity of the chamber is recommended and, if excessive outgassing or vapour evolution occurs, this figure should be doubled. A gas-ballast rotary pump is used for backing the oil vapour pump. For a 12 inch diameter bell-jar, a vapour pump of a speed able to provide about 60 litre per sec at the vessel, with a backing pump of displacement 30 litre per min, is suitable. As in the system described in section 3.6, a by-pass or roughing line is provided. The connections are now made by $\frac{1}{2}$ inch inside diameter, copper tubing, with soldered joints and metal vacuum valves, of which the Saunder's type (Fig. 76*a*) is convenient and economical. The main isolation valve between the oil diffusion pump intake port and the base plate to the bell-jar (Fig. 76*b*) has a neoprene or Viton O-ring seal. The demountable seal between the bell-jar and the ground base

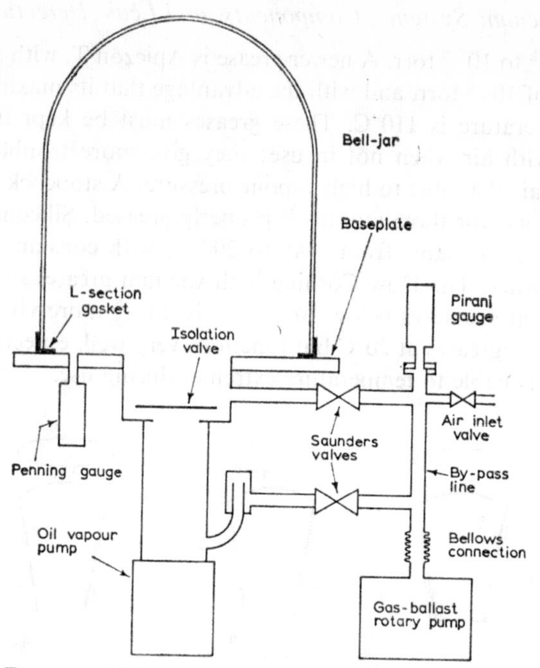

Fig. 75. Demountable laboratory vacuum system with oil vapour pump.

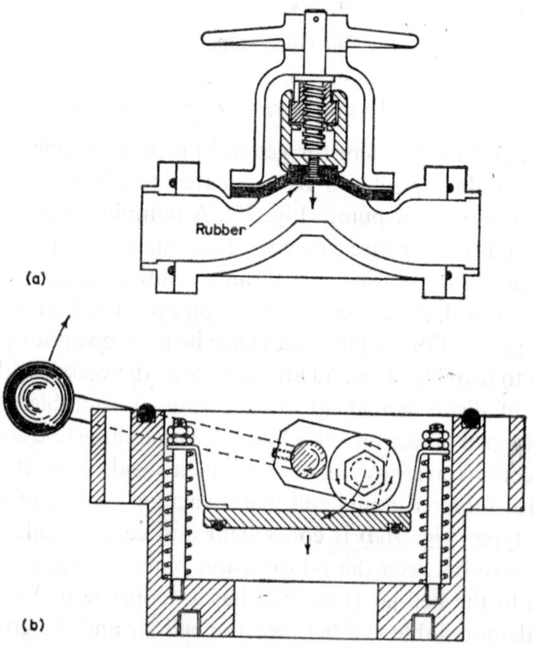

Fig. 76. (*a*) A Saunder's valve. (*b*) An isolation valve.

plate is either by means of an L-section neoprene or Viton gasket, or the bell-jar rim has a ground flange which seats on an O-ring in a groove cut in the base plate. A Penning gauge is convenient for recording the pressure attained in the bell-jar, and a Pirani gauge to record the backing pressure is an advantage.

The backing pump must be chosen to cope adequately with the throughput of the gas from the vapour pump, especially in a dynamic system in which there is a continual evolution of gas due to the process undertaken in the chamber. If the throughput is Q torr litre per sec, p_i is the intake pressure at the vapour pump, and p_b the backing pressure, then

$$Q = p_i S_v = p_b S_b$$

where S_v is the speed at the pressure p_i of the vapour pump, and S_b the speed at p_b of the backing pump, both in litre per sec. Therefore,

$$S_b = \frac{p_i}{p_b} S_v$$

In practice, it is best, if possible, to choose a value of S_b two or three times this calculated value.

The **O-ring gasket** is a valuable aid in vacuum technique, enabling quickly demountable seals to be made. Of the several elastomer materials available, nitrile rubbers, neoprene, and Viton are the most frequently used. The first mentioned are oil resistant and have been widely used; the more expensive Viton is gaining favour, as this fluorocarbon material is resistant to temperatures up to 250°C and has much lower gas permeability than neoprene. Recommended seatings and groove dimensions for O-rings are shown in Fig. 77 and Table 3.2 in conjunction with the dimensions of the O-rings themselves given in Table 3.1. Baking of O-rings and other elastomer gaskets in air or preferably in a vacuum is good practice, to reduce their outgassing in use, but is not necessary for the type of system described in Fig. 75. Light greasing with Apiezon or silicone greases facilitates sealing, especially when the smaller sizes are used.

Quickly assembled vacuum unions and pipe-lines based on O-ring seals are a feature of the products of the vacuum equipment suppliers used to facilitate vacuum system construction. As an example of a convenient pipe-line assembly for rapid erection of the connections for a system like that of Fig. 75, see Fig. 78. **Lead-in terminals** for electrical supplies into a vacuum chamber are commercially available in a wide variety. Two convenient designs, one based on an

TABLE 3.1
O-rings

Edwards VOR No.	*British* Standard No.*	*Inside diameter D (inch)*	*Section diameter d (inch)*	*Edwards VOR No.*	*British Standard No.*	*Inside diameter D (inch)*	*Section diameter d (inch)*
101	006	0·114	0·070	179	337	2·975	0·210
104	009	0·208	0·070	182	339	3·225	0·210
105	010	0·239	0·070	184	341	3·475	0·210
107	011	0·301	0·070	190	345	3·975	0·210
110	012	0·364	0·070	194	348	4·350	0·210
116		0·487	0·103	196	425	4·475	0·275
118	113	0·549	0·103	205	429	4·975	0·275
120	114	0·612	0·103	214	433	5·475	0·275
122	115	0·674	0·103	227	439	6·475	0·275
124	116	0·734	0·103	232	441	6·975	0·275
129	212	0·859	0·139	236	443	7·475	0·275
133	214	0·984	0·139	240	445	7·975	0·275
135	215	1·046	0·139	242	446	8·475	0·275
136	216	1·109	0·139	244	447	8·975	0·275
138	218	1·234	0·139	246	448	9·475	0·275
140	220	1·359	0·139	248	449	9·975	0·275
142	222	1·484	0·139	252	451	10·975	0·275
148	327	1·725	0·210	256	453	11·975	0·275
154	329	1·975	0·210	258	455	12·975	0·275
160	331	2·225	0·210	260	457	13·975	0·275
166	333	2·475	0·210	262	459	14·975	0·275

* Same as U.S.A. MS29513 dash number except that B.S. numbers 006 to 028 are referred to as 6 to 28 in U.S.A. standards.

TABLE 3.2
Dovetail groove dimensions [Figs. 77*b* and *c*]

Diameter d of section of O-ring	*A*	*B*	*C*
0·070	0·059	0·095	0·049
0·103	0·088	0·142	0·074
0·139	0·121	0·194	0·100
0·210	0·185	0·293	0·149
0·275	0·243	0·391	0·203

Trapezium groove dimensions [Fig. 77*d*]

Diameter d of section of O-ring	*W*	*X*	*Y*
0·070	0·065	0·046	0·086
0·103	0·097	0·063	0·123
0·139	0·132	0·084	0·166
0·210	0·200	0·124	0·248
0·275	0·264	0·161	0·323

All dimensions are in inch.

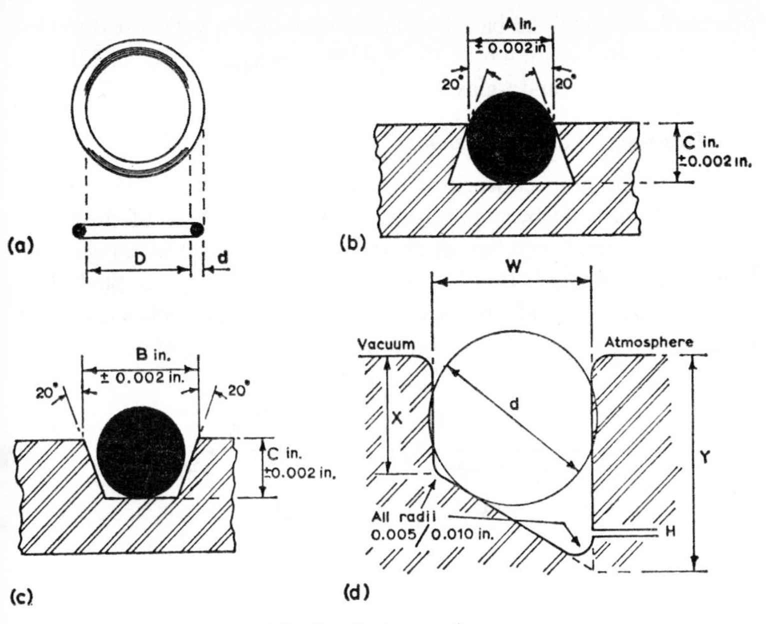

Fig. 77. O-ring seatings.

O-ring and glass insulation and the other on a ceramic-to-metal seal, are illustrated in Fig. 79. Lead-in electrodes are also frequently based on glazed porcelain, with an appropriate band of deposited platinum to which a soldered joint is made (see p. 252). Where bake-out is to be

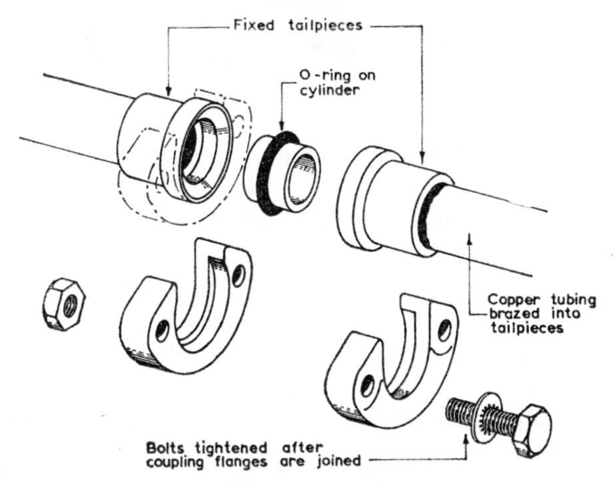

Fig. 78. A rapidly assembled pipe-line using standard vacuum unions (Edwards High Vacuum Ltd.).

practised, they may be based on an alumina body, with a seal to the metal made by titanium hydride or zirconium hydride (see p. 221), or on the use of Kovar-to-glass seals (see p. 239).

3.8. *Ultra-high Vacuum Systems*

Following the pioneer work of Alpert [93] in 1950 to 1953, a great many papers have been published on the production of pressures below 10^{-8} torr. From this work, the present position with regard to laboratory systems is that two basic types are widely adopted. The first is a development from the conventional arrangement of a vapour diffusion pump backed by an oil-sealed rotary pump, where a liquid-

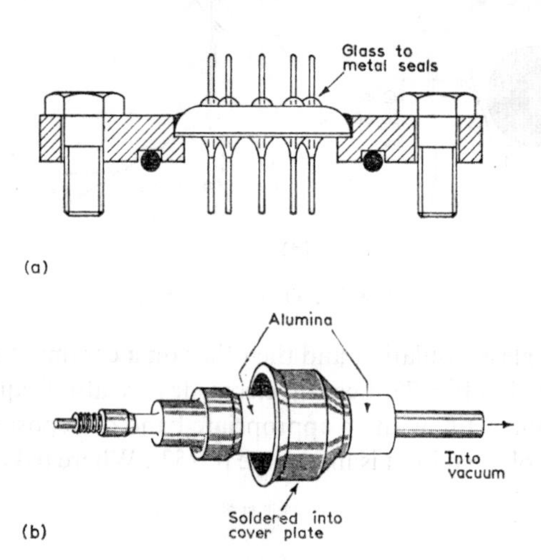

Fig. 79. Lead-in terminals.

nitrogen trap is used between the diffusion pump and the chamber; the second employs a cold-cathode getter-ion (sputter-ion) pump with backing by a sorption pump, usually based on molecular sieve materials. The most frequently used gauge to measure the low pressures obtained is the Bayard-Alpert gauge (section 2.7).

The ultimate pressure p_u attainable in a pumped chamber will be the result of a balance between the rate of production of gas within the chamber, Q, and the effective speed of pumping, S_E. The amount of gas removed per unit time will be p_uS_E at the ultimate, and clearly

$$p_uS_E = Q \tag{3.20}$$

Q will be due to gas entering the chamber through actual leaks, virtual leaks caused by gas and vapour evolution from the chamber walls and materials inside, added to which is any gas or vapour entering the chamber from the pump itself. From equation (3.20), it follows that an ultimate pressure of 10^{-n} torr can be obtained with an effective pumping speed of S_E litre per sec, only if Q does not exceed

$$Q = 10^{-n} S_E \text{ torr litre sec}^{-1}$$

With limited pumping speed, Q must be very small if n is to be greater than 8 and, as is often the requirement, equal to 10 or 11.

With S_E at, say, 5 litre per sec, to produce a pressure of 10^{-9} torr demands a value of Q not exceeding 5×10^{-9} torr litre per sec. The achievement of such low leakage rates offers some difficulty as may be seen from a simple calculation. Suppose the chamber concerned has a wall thickness of 2 mm and that there exists in it a uniform bore hole of diameter d. Common sense shows that d will be much less than 2 mm if 10^{-9} torr is to be maintained. The hole can therefore be regarded as a cylindrical tube of length much greater than its diameter. At one end of this hole exists the atmosphere, at a pressure of 760 torr; at the other end, a pressure to be maintained at 10^{-9} torr. The flow of gas through the hole is viscous at its entrance and certainly molecular at its exit. To simplify the calculation, it can be assumed that the hole diameter will need to be very small, and that the pressure over most of its length will be low enough for the m.f.p. to exceed the diameter d. The equation (3.14) for the molecular conductance U of a long narrow tube can then be applied. Hence,

$$U = \frac{12{\cdot}1 d^3}{l} = \frac{12{\cdot}1 d^3}{0{\cdot}2}$$

The throughput Q of air is $760U$. This must not exceed 5×10^{-9} torr litre per sec. It follows that the hole diameter d must not exceed a value given in cm by

$$\frac{5 \times 10^{-9}}{760} = \frac{12{\cdot}1 d^3}{0{\cdot}2}$$

The maximum tolerable hole diameter in this case is consequently

$$d = \sqrt[3]{\left(\frac{10^{-9}}{760 \times 12{\cdot}1}\right)} = 4{\cdot}8 \times 10^{-5} \text{ cm}$$

It is clear that not only will the chamber wall have to be free from very tiny pores and cracks, but also any seals will need to be very carefully made.

The problem of producing an ultra-high vacuum, and then maintaining it within an isolated volume of, say, 1 litre, may be said to be aggravated by the fact that the volume occupied by 1 mole of material in the gaseous state is as large as 22·4 litre at s.t.p. Suppose a speck of mass m gram, of solid material of molecular weight 20, exists initially within the isolated chamber and that this material vaporizes. Within a time, depending on the rate of evaporation, the pressure exerted within the chamber of volume 1 litre will be $760 \times 22{\cdot}4m/20$ torr, presuming that m is so small that this pressure is less than the saturated vapour pressure. If this pressure is to be only 10^{-9} torr, it follows that the maximum amount of vaporizable material allowable within the isolated vessel is

$$m = \frac{20 \times 10^{-9}}{760 \times 22{\cdot}4} = 1{\cdot}2 \times 10^{-12} \text{ gram}$$

which is about one-millionth of the minimum mass weighable by a sensitive chemical microbalance.

This rough calculation is misleading in that such minute amounts of material would usually become bonded to the chamber walls and not pass readily into the gaseous phase within the space. However, the need for scrupulous cleanliness of apparatus made from specific bakeable materials is stressed. Even a finger-print can be disastrous!

To achieve ultra-high vacua, it is thus necessary to provide a chamber which is scrupulously clean, free of leaks, and rigorously degassed by bake-out during pumping. Furthermore, materials within the chamber have to be degassed and also seals, gaskets, vacuum valves, and any traps used between the chamber and the diffusion pump or the getter-ion pump. The chamber is usually made of either borosilicate glass or a stainless steel selected for its resistance to corrosion, workability, and the facility with which it can be welded or brazed (p. 258). In the case of glass, seals are made by direct glassblowing between glass and glass, or matched glass and metal (p. 238). It is worth noting that the usual Tesla coil means of indicating pinholes is often inadequate here. With stainless steel chambers, it is best to electropolish (p. 226) the inner walls, and seals are made by argon-arc welding or preferably by vacuum brazing or, if to be demountable, by the use of metal gaskets.

Bakeable **ultra-high vacuum valves** follow the pioneer design of Alpert [94], who devised an all-metal valve, without grease, which could be degassed by bake-out at 400°C. This valve can also be used as an excellent 'needle valve' for the metering of pure gases into vacuum systems. A more recent version due to Baker [95] (Fig. 80) depends on making a closure by driving an annealed, flat, copper disk, forming the valve plate, against a stainless steel knife-edge, forming the valve seat, machined in the stainless steel valve block. To close the valve, the shaft joined to the valve plate is driven forward by a simple screw, $\frac{5}{16}$ inch B.S.F. in the case of a valve of 2·5 cm aperture, which is lubricated by molybdenum disulphide. To provide a flexible vacuum-tight seal between this driving shaft and the valve body, one end of a stainless steel bellows is argon-arc welded to the

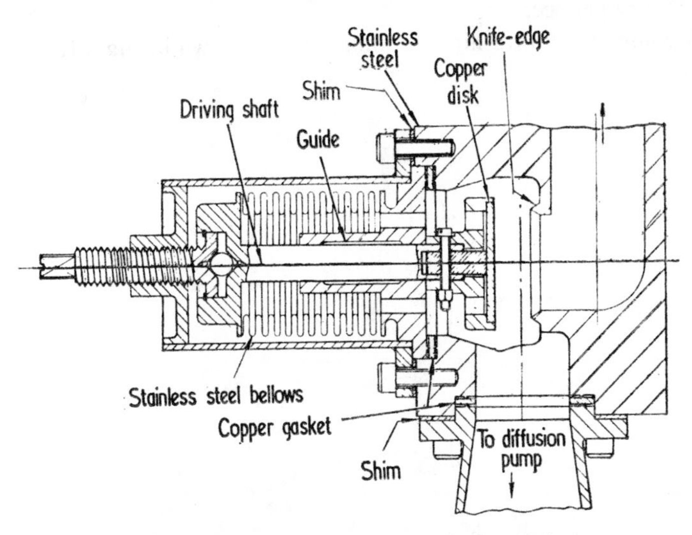

Fig. 80. Metal ultra-high vacuum valve (Baker).

shaft and the other end to a guide, which is sealed to the valve body by a copper gasket between knife edges. A shim is used to limit the penetration of these knife-edges in the body seals to 0·007 inch, when the Allen screws holding the drive assembly to the valve body are tightened. When the driving shaft forces the valve plate against the seat, the body seal is then unaffected, the closing force being exerted against the Allen screws.

The knife-edges, employed both at the stainless steel valve seat and at copper seals within the valve body, are 60°. After machining, they are first lapped with a brass lap using 15 micron silicon carbide, followed by a second brass lap using 3 micron diamond powder, to produce a perfectly polished surface, 0·003 to 0·010 inch in width at the top of the knife-edge. Inspection for adequate finish involves that no scratches are visible by a microscope providing linear magnification of × 50.

The copper sealing disk and gaskets used were machined from OFHC (oxygen-free, high conductivity) copper and annealed in hydrogen before assembly.

This valve can be baked to 400°C in both the open and closed positions on the vacuum system. The version with a 2·5 cm aperture requires a normal closure torque of 2 to 2·5 lb foot, has an open conductance of 15 litre per second and a closed conductance of less than 10^{-14} litre per sec.

A commercial ultra-high vacuum valve is shown in Fig. 81.

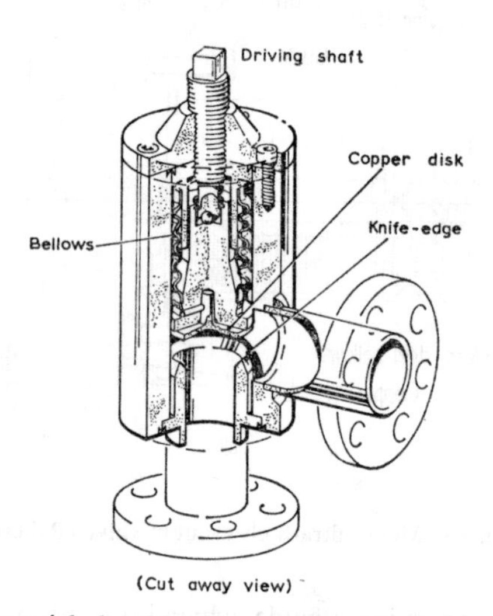

Fig. 81. A commercial ultra-high vacuum valve (Vacuum Generators Ltd.).

These metal valves are chiefly used with all-metal vacuum systems, but they can also be used as closure valves and gas-metering devices in glass systems. It is then important to clamp the valve firmly to a rigid support, usually the thick 'Sindanyo' floor of the system bake-out oven, and to provide a flexible loop in the glass tubing between the glass system and the glass-to-metal seal joining to the valve itself. Otherwise, the torque exerted on the valve driving mechanism on closure may fracture the glass.

An economical and simple alternative to the metal valve with a glass system is a magnetically-operated, greaseless, glass valve (Fig. 82), where the necessary closure is effected between two, mating, hemispherical, ground glass joints. Vacuum grease is *not* used; indeed, greases cannot be tolerated within an ultra-high vacuum system (except on the backing side) because of their outgassing and the necessity for bake-out. When closed, the seal between the ungreased glass hemispheres is not of minute conductance, but it is small enough for the ultra-high vacuum produced within the isolated glass chamber to be maintained, provided the pumps are kept running.

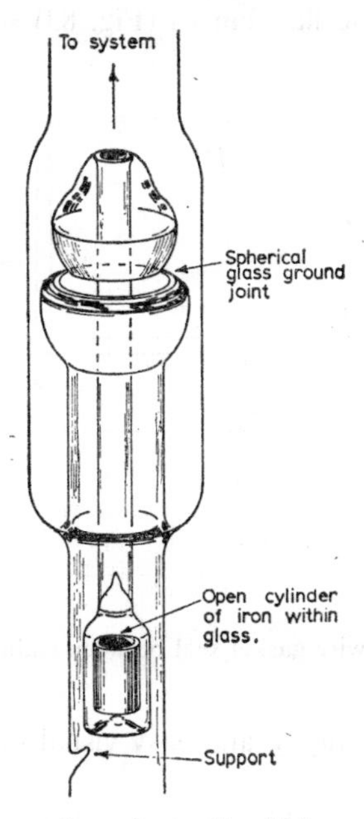

Fig. 82. A greaseless, glass, ultra-high vacuum valve.

Metal gaskets between stainless steel flanges and components which are bakeable to 450°C have been made of copper, aluminium, gold, and other metals. Aluminium gaskets of the type introduced by Holden, Holland, and Laurenson [96], and shear gaskets of the Brymner and Steckelmacher [97] pattern, using a flat sheet gasket of copper, stainless steel, or nickel between specially bevelled edges, have both been used successfully, but reliability over long periods where frequent bake-out is practised undoubtedly favours the use of gold, in our experience. Several methods of using gold gaskets have been described; the originator appears to have been Hickam [98] in 1949 followed by Grove [99] in 1958. More recently, there has been a desirable increase in favour of the idea that the machining of the stainless steel flanges should be simple and not involve awkward shaping, and further that these flanges should be 'sexless', i.e. they should not have male and female components of different geometry, and so that they can be readily interchanged. A type developed by K. J. Close (private communication) is similar to that favoured by Edwards High Vacuum Ltd. This firm employs straightforward flat flanges; the development is simply to provide polished annular surfaces proud above the flat flanges (Fig. 83) so that mating can be

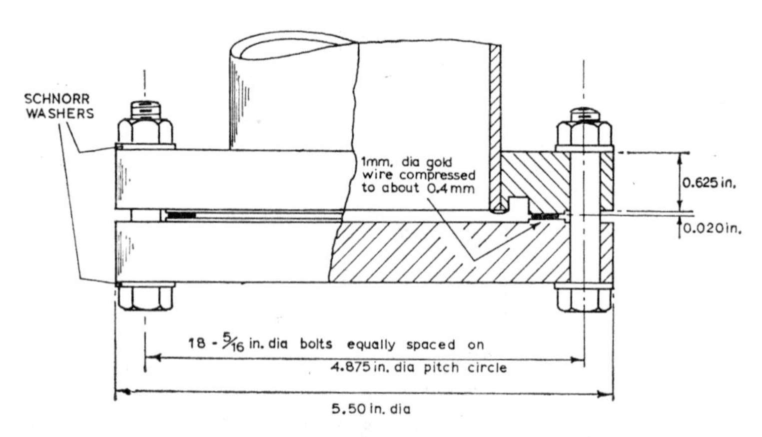

Fig. 83. A gold-wire gasket seal between stainless steel flanges.

monitored by feeler gauges, and easy visual inspection of the compressed gold wire is possible.

To ensure success in making such a gasket seal, attention is drawn to the following points.

(*a*) Use 24 carat gold wire of 0·02 inch (0·5 mm) diameter; (0·04 inch (1 mm) diameter wire is also used, especially for large diameter flanges).

(*b*) Form the wire around a suitable cylinder, remove it, and lay it flat upon a clean block of refractory material with the carefully square-cut ends of the loop adjacent. Fuse these ends together by playing upon them a fine oxygen-gas blow-pipe flame. If this is done carefully, it is difficult to see the joint with the naked eye.

(*c*) Anneal the gasket by heating it all round to red heat in an air-gas blow-pipe flame. Movement of the wire will occur as the strain is removed. Reshape the gasket around a cylinder, if necessary, but ensure that it is not touched by hand between the final annealing and assembly between the flanges. This means careful manipulation with tweezers or the wearing of clean finger-stalls.

(*d*) The stainless steel flanges, between which the gold gasket is to be placed, are machined and ground to the profile shown in Fig. 83. These flanges are best 0·625 inch or more thick, and 1·0 inch is desirable for flange diameters exceeding 6 inch. The proud annular surfaces to be mated with the gasket sandwiched between them are about 1 cm (0·4 inch) wide. These surfaces have to be machined first (lathe-work is capable of 50 to 100 microinch finish) and then ground and polished to a final finish of about 16 microinch r.m.s. Brass laps, with first 15 micron silicon carbide followed by 3 micron diamond powder, are ideal for grinding and polishing, but a less stringent technique giving a final finish of about 30 microinch is probably adequate. Few laboratories have the equipment needed to measure fine surface irregularities; experience enables the naked eye to detect a good enough polished finish, free from scratches, but it is better to inspect finally with a low-power microscope.

(*e*) The thick flanges have to be very firmly bolted together by stainless steel bolts, provided with Schnorr washers to ensure that compression is maintained during bake-out under vacuum to 450°C. Usually, a bolt every $\frac{1}{2}$ inch of the flange circumference is employed, but some experience indicates that fewer bolts are adequate with larger flanges of 1 inch thick stainless steel.

(*f*) The bolts are tightened up in a sequence ensuring uniform compression of the gold-wire gasket. The diameter of the wire is reduced from the initial 0·020 inch to 0·007 inch, or 0·040 inch to 0·016 inch, during the compression.

(*g*) The assembled junction on the vacuum system is then baked out under vacuum to about 450°C. During bake-out, the gold adheres to the stainless steel to provide an excellent seal. Gold with its melting point of 1,063°C, low vapour pressure and resistance to oxidation is ideal; its thermal expansion coefficient of $14{\cdot}3 \times 10^{-6}$ per degC is less than that of 18/8 stainless steel at 17×10^{-6} per degC, but this does not appear to offer difficulty. Silver wire has also been used in place of gold.

The ultimate pressure provided by a vapour diffusion pump in a chamber with insignificant leakage has been the subject of numerous studies. The use of the **vapour diffusion pump to obtain ultra-high vacua** stems from the pioneer investigations of Venema [100]. He lists four sources of gas flow through the pump mouth in the direction towards the chamber:

(*i*) the vapour of the pump fluid and the decomposition products of this fluid;
(*ii*) back diffusion of gas from the backing side of the pump to the intake port;
(*iii*) gas production from the walls of the pump near its mouth;
(*iv*) gas dissolved in the cold condensed pump fluid which is transported again to the top-stage jet.

To minimize these contributions of the pump to gas in the chamber, Venema chose mercury as the pump fluid instead of oil, as it cannot decompose, and he reduced its vapour pressure in the chamber to exceedingly low values by three, glass, liquid-nitrogen traps in series, of the type shown in Fig. 22(*d*), above the glass diffusion pump. To minimize the factor (*ii*), the backing pressure to this glass pump was maintained at below 10^{-5} torr by a metal, mercury diffusion pump as an intermediary between the glass one and the oil-sealed rotary backing pump. Gas production from the pump walls in the vicinity of its mouth (factor *iii*) was reduced by arranging degassing of the upper part of the glass pump by bake-out. By adopting a long and specific bake-out procedure, he attained ultimate pressures of 10^{-11} to 10^{-12} torr.

Subsequent to Venema's work, it has become general practice to use mercury and also oil diffusion pumps, usually of metal, and surmounted by a liquid-nitrogen-filled cold trap of stainless steel (Fig. 84), to attain ultra-high vacua. Controversy exists as to whether the intermediate vapour pump is needed between the main vapour diffusion pump and the mechanical backing pump. Experience shows that to attain ultimates of 10^{-10} torr, it is unnecessary provided a matched two-stage oil-sealed rotary pump is used for backing, but it is probably an advantage if lower pressures are required. However, a cold trap between the mechanical pump and the diffusion pump to prevent oil vapour from the former entering the latter is desirable. Again, the use of oil diffusion pumps has the attraction, compared with mercury, that any fluid which does by mischance get above the cold trap is less troublesome. Silicone 704 and especially silicone 705 and Convalex 10, with vapour pressures at room temperature of about 10^{-10} torr, have therefore been much used in recent work. It is also valuable to be able to bake-out the cold trap before filling, and there are advantages gained by using two cold traps in series above the diffusion pump with the top one bakeable.

An ultra-high vacuum system which has been consistently used over many months to obtain ultimates of 5×10^{-10} torr and below is shown in Fig. 84. A single cold trap is used which is not baked. A 2 inch port diameter, oil diffusion pump filled with silicone 704 fluid is surmounted by a water-cooled cup-shaped baffle of the type shown in Fig. 18(*c*), above which is a stainless steel liquid-nitrogen-filled cold trap (Fig. 23*a*). A 1 inch orifice diameter, bakeable, ultra-high vacuum, metal closure valve connects the intake to the cold trap to the

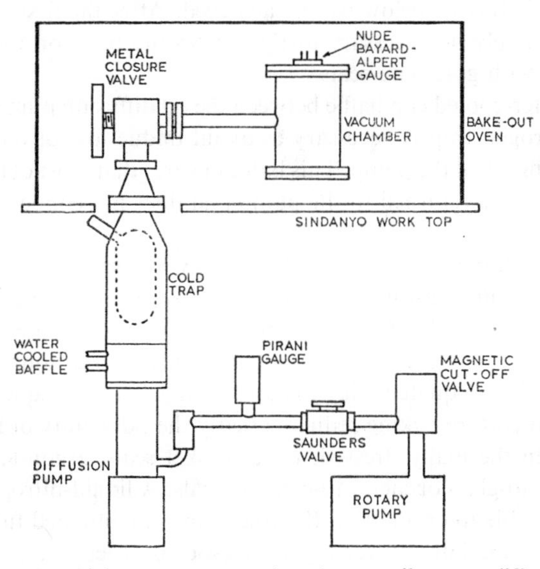

Fig. 84. An ultra-high vacuum system based on an oil vapour diffusion pump.

chamber, which is of argon-arc-welded stainless steel with electropolished inner walls. The oil diffusion pump is backed by a matched two-stage oil-sealed mechanical rotary pump with gas ballast. The O-ring seals, between the top flange of the oil diffusion pump and the baffle, and between the baffle and the base of the cold trap, are of indium wire or Viton, but the former is preferred. The other metal gaskets, between the top of the cold trap and the metal closure valve, and between this valve and the chamber, are of gold wire. A metal-to-glass seal is necessary if a glass chamber is used.

The procedure in using this system is first to establish a pressure of about 10^{-6} torr in the chamber, by running the pumps in the normal manner but without filling the cold trap. An oven is then arranged around the chamber, with Bayard-Alpert gauge attached and including the open, metal closure valve, and bake-out is commenced with the temperature raised gradually to 400°C. After initial baking for about 1 hour, the cold trap is filled with liquid nitrogen, and bake-out continued for another 3 hour for a chamber of volume not exceeding 2 to 3 litre. After bake-out, the Bayard-Alpert gauge is switched on and its electrodes degassed by heating to about 900°C (bright red) for at least 30 min. A further bake-out in the oven is then undertaken. On subsequent cooling to room temperature, a pressure of 5×10^{-10} torr or below will be achieved. After the first bake-out, it will probably be necessary to tighten up the bolts on the flanges furnished with gold wire seals.

The water-cooled cup baffle between the oil diffusion pump and the liquid-nitrogen trap is necessary to avoid undue loss of oil on prolonged running of the pump. This baffle ensures that most of the back-streaming oil is returned to the pump and does not condense on the cold trap.

Such a comparatively simple system is very useful for laboratory work at pressures down to 5×10^{-10} torr. After first commissioning with a bake-out at 400°C, the chamber should be kept under vacuum as far as is possible when bake-outs for only 1 to 2 hour at about 250°C will be adequate in day to day running. The subsequent use of lower temperature baking reduces greatly the possibility of leaks developing in thermally-stressed, metal gasket seals, argon-arc welds, and metal-to-glass or metal-to-ceramic seals. A liquid-nitrogen leveller is advisable to ensure that the cold trap is maintained full during operation, especially if over-night bake-out is practised.

Any metal electrodes or other parts introduced into the chamber for experimental work must be rigorously cleaned and degassed. Degassing of the clean metal in trichlorethylene is desirable, followed by stoving in first a hydrogen furnace and then a vacuum furnace, or at least one of these, before the part is introduced into the chamber to be evacuated.

Indium gaskets referred to above are useful for making seals in a system where an ultimate pressure of about 10^{-8} torr is adequate, but bake-out cannot be practised on such seals except at moderate temperatures up to 140°C, because the melting point of indium is 156°C. They are very easily made and desirably free of organic materials. Indium wire of $\frac{1}{32}$ or $\frac{1}{16}$ inch diameter is simply formed into a ring with an overlap between the ends. This O-ring of indium is then compressed by bolting between flat flanges on copper or stainless steel piping or components, until its thickness is reduced to 0·005 to 0·01 inch. The overlapping ends of the indium O-ring will cold-weld readily during this compression.

The logical extension of Venema's original work with glass systems to the use of metal is exemplified by the interesting design of Power, Dennis and Csernatony [101], which is capable of ultimates of 10^{-12} torr and below. To ensure complete freedom from organic materials, a mercury diffusion pump is used backed by a second mercury vapour pump, capable of operating with backing pressures of 30 torr, which in turn is backed by a sorption pump. Surmounting the main mercury diffusion pump mouth are first a chevron baffle cooled thermoelectrically, and then a Z-type baffle within the base of a liquid-nitrogen trap. Furthermore, to ensure the utmost freedom from gasket leakage, the metal gasket, between the mercury diffusion pump and the chevron, Z-baffle, liquid-nitrogen trap unit, is well below the top first-stage jet of the mercury diffusion pump and between the two water-jackets. This gasket is thus below the baking region, which includes the top of the diffusion pump, both baffles and the cold trap. Thus, the seal made from silicon-aluminium wire is not only freed from metal gasket leaks that may arise on bake-out, but it is also subject to less stringent leak-free requirements.

The 'dry' vacuum obtained by the cold-cathode getter-ion pump without the use of cold traps is undoubtedly attractive in setting up an ultra-high vacuum system. Indeed, some laboratories engaged on investigations under ultra-high vacua employ this type of pump exclusively, on the basis that it is less troublesome than the vapour diffusion type, even though the capital cost is greater and the ability to handle considerable gas loads is more limited.

An ultra-high vacuum system based on a cold-cathode getter-ion pump employs two sorption pumps of the molecular sieve type for backing. Alternatively, two getter-ion pumps may be used and a single sorption pump (Fig. 85). In general, the getter-ion pump will produce a pressure of about 10^{-8} torr in a baked chamber, but to enable it to produce better vacua, in the 10^{-10} torr region, it is essential to bake the pump itself. During the bake-out of the cold-cathode or sputter-ion pump, about 97% of the gas released from its titanium electrodes and envelope is hydrogen; the remainder is helium, methane, argon, nitrogen, water vapour, and carbon dioxide (Rivière and

Allinson [102]). Moreover, after pumping and switching-off the H.T. to the electrodes, gas – chiefly hydrogen – will be evolved even at room temperature. Consequently, during bake-out of the pump to achieve subsequently ultra-high vacua, it is essential to provide means of removing this gas.

Good practice is represented by the system of Rivière and Allinson, which utilizes two getter-ion pumps, each of speed 40 litre per sec. The sorption pump chilled with liquid nitrogen reduces the pressure initially to about $2{\cdot}5 \times 10^{-2}$ torr, as recorded by a Pirani gauge. One

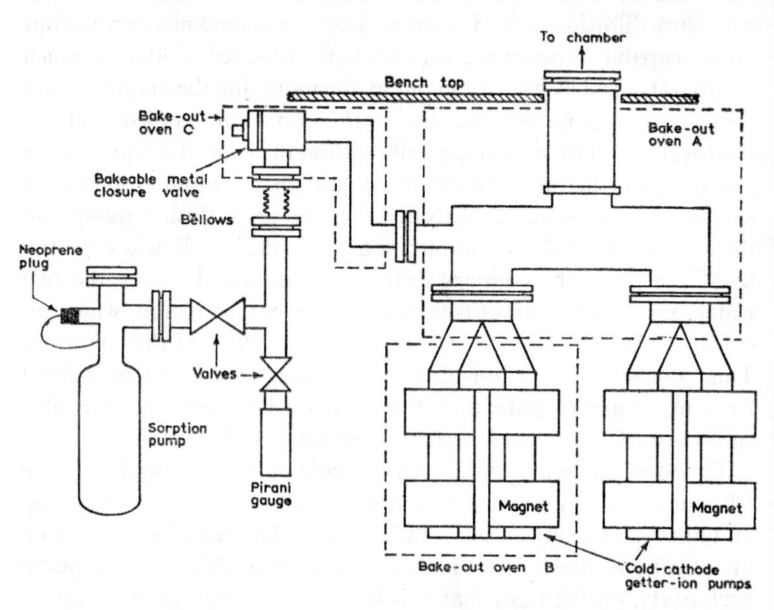

Fig. 85. Ultra-high vacuum system based on a cold-cathode getter-ion pump.

of the getter-ion pumps is then switched on. If this pump has not been baked, the initial heavy discharge current causes outgassing of its electrodes, so the pressure rises. The pump has then to be switched-off and switched-on again after a delay, until it begins to pump satisfactorily. The diaphragm valve is then closed to isolate the sorption pump when the pressure drops to about 10^{-7} torr. Three bake-out ovens are used: A to bake the manifold and chamber; B is arranged around either one of the getter-ion pumps; C is used to heat the ultra-high vacuum, metal closure valve and connecting tubing. A and C could readily be combined in one oven.

The procedure is to bake at 450°C the chamber, manifold and metal closure valve by means of ovens A and C, and meanwhile bake the getter-ions pumps alternatively with oven B. Whilst one of these pumps is being baked, its magnet is removed, whilst the other pump is switched on. After two days of pumping and bake-out in this way and subsequent cooling to room temperature a pressure is obtained approaching 10^{-11} torr, i.e. below the X-ray limit of the thoroughly degassed nude Bayard-Alpert gauge.

The procedure with two sorption pumps and a single cold-cathode getter-ion pump is to pump the system first by chilling one of the sorption pumps with liquid nitrogen. A pressure of about 2×10^{-2} torr is attained, which also prevails in the second sorption pump. The first sorption pump is then isolated by a valve, and the second sorption pump is liquid-nitrogen chilled. The pressure will then fall to about 10^{-4} torr, when the chamber and the getter-ion pump is baked to 450°C. Gas evolved from the getter-ion pump is then largely taken up by the sorption pump. After bake-out, the system is cooled, the second sorption pump is isolated by a valve and the degassed getter-ion pump is switched on.

The commercial cold-cathode getter-ion pump systems of Varian Associates now include a titanium sublimation pump as an auxiliary or intermediate between the getter-ion pump and the sorption pump. This not only ensures low starting pressures for the getter-ion pump, but also provides extra pumping speed for getterable gases in the system.

3.9. *Vacuum Systems for Gauge Calibration*

There are two main ways of calibrating a vacuum gauge so that its output reading is known at various pressures for a known gas. The first is to compare it with a McLeod gauge, considered to be an absolute instrument, or a mercury or oil-filled manometer, if the pressure range to be covered exceeds 10^{-1} torr; the second way is to use the calculated molecular flow through an orifice of known geometry as the absolute standard. The McLeod gauge is not a suitable instrument for gauge calibration at pressures below 10^{-4} torr, for the reasons discussed in section 2.3, though extension to 10^{-5} torr is possible if elaborate precautions are taken. The second method, of using an orifice, is valid below 10^{-4} torr where molecular conductance prevails. It is therefore a good choice for calibrating an ionization gauge, but there are difficulties in providing adequate pumping speed at below 10^{-8} torr; then a series of orifices is necessary with large pumps and prolonged bake-out procedures, leading to an expensive set-up.

A system for calibrating a gauge against a McLeod gauge is a static one for the pressure range from 10^{-1} to 10^{-4} torr. It demands that the McLeod gauge be constructed taking into account the precautions given in section 2.3; that mercury vapour from the McLeod gauge is excluded from the gauge head under test by the use of a liquid-nitrogen-filled cold trap; that the McLeod gauge, together with other parts of the system at low pressure, is baked-out at 400°C; and, finally, that a gas-metering system is included to admit the dry, known gas to the

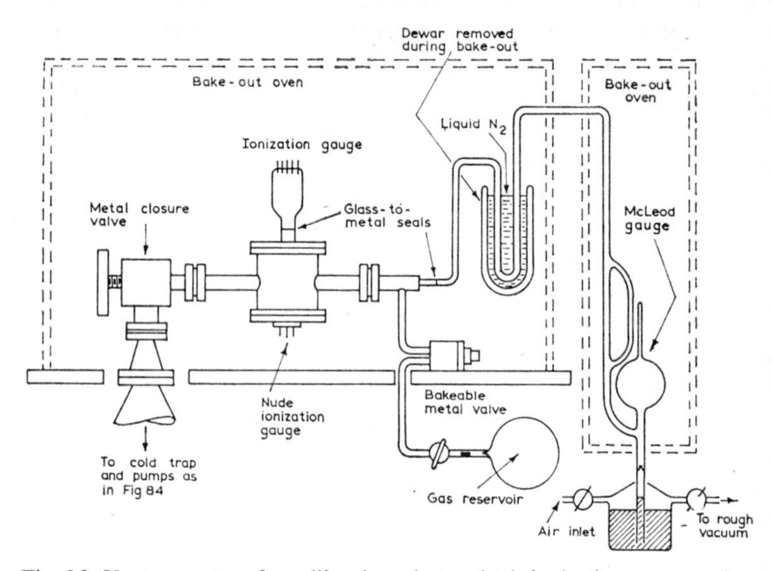

Fig. 86. Vacuum system for calibrating a hot-cathode ionization gauge against a McLeod gauge.

initially evacuated system. It is clear, therefore, that some of the techniques practised in ultra-high vacuum systems are used, such as the employment of bakeable metal taps and metal gaskets. For student's use, this leads to an expensive construction requiring careful operation. In a college, therefore, a simpler, glass system employing greased valves and limited bake-out of only the chamber to which the gauge under test is connected can be used; it will be instructive but inaccurate, except at pressures in excess of about 5×10^{-3} torr.

The rather elaborate system described here (Fig. 86) is recommended; simplification of it in constructing an all-glass system for student's use can readily be accomplished. The practice is to obtain first an ultimate pressure of 10^{-8} torr or below, isolate the system from the pumps by the metal valve closure, ensure that the subsequent pressure rise is not to above 10^{-6} torr in 1 hour, re-establish the ultimate, isolate the pumps again, and admit the gas via a bakeable metal valve from a reservoir of spectroscopically-pure gas, to give various pressures in the range from 10^{-4} to 10^{-1} torr as indicated by the McLeod gauge; within the isolated static system note the reading of the ionization gauge. To ensure that this gauge is not pumping significantly, its electron current should be 100 μA or less. The advantage of this static procedure is then that the pressure is the same throughout the isolated volume.

To be able to satisfy the main requirement, that the pressure in the isolated system rises at a rate less than 10^{-6} torr per hour, the use of an oil vapour diffusion pump surmounted by a stainless steel liquid-nitrogen cold trap and a bakeable metal closure valve of 1 inch orifice, with backing by an oil-sealed rotary pump on the lines of the system shown in Fig. 84, is recommended. The chamber to which the ionization gauge is attached can be glass connected via a glass-to-metal seal to the metal closure valve, or stainless steel with, say, one gauge connected via a glass-to-metal seal and the other a nude gauge mounted on a stainless steel cover plate with appropriate lead-ins and gold-wire sealing. Between the chamber, to which the ionization gauge is connected, and the McLeod gauge, it is essential to have a cold trap. This is best in the form of a symmetrical U-tube, to avoid pressure gradients across it on cooling with liquid nitrogen. An asymmetrical trap of the type shown in Fig. 22(*b*) would have a temperature difference between inlet and outlet on cooling, which would introduce an unknown pressure difference due to thermal transpiration in the gas. The ovens shown in Fig. 86 are arranged to bake-out the system (except the pumps) including the McLeod gauge as far down as the tube leading to its mercury reservoir.

The ionization gauge must be thoroughly degassed by electron bombardment or eddy-current heating before it is calibrated.

The **orifice method of calibrating an ionization gauge** is a dynamic one, in that a measured gas throughput is passed through the vacuum system during the test procedure. In use for many years at the laboratories of Associated Electrical Industries Ltd., Manchester (Leck [103]), this technique has gained much favour recently as the hitherto unsuspected errors of the McLeod gauge have become more apparent. Normand [104], Roehrig and Simons [105], and Feakes and Torney [106] have been recent exponents of the technique, and it may well become accepted by national and international standards organizations.

The principle adopted (Fig. 87*a*) is to connect the gauge G_1 to be calibrated within or to a measuring chamber A, which is separated by an orifice O of calculated molecular conductance from a second chamber B, to which a vapour diffusion pump is directly attached. A second, similar gauge, G_2, is attached to chamber B or, alternatively, by a simple arrangement of bakeable metal closure valves, one gauge only, G_1, is used and connected at will to either A or B.

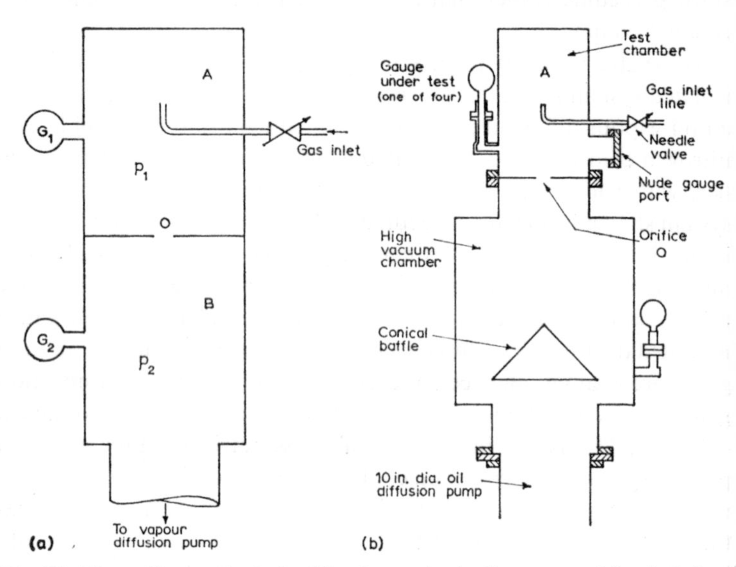

Fig. 87. The orifice method of calibrating an ionization gauge: (*a*) principle of the method; (*b*) dynamic vacuum system used by Normand.

If p_1 is the gas pressure above the diaphragm in chamber A, and p_2 is the gas pressure below the diaphragm in chamber B, whilst U is the molecular conductance of the orifice of area A, as calculated from equation (3.7), it follows that

$$Q = U(p_1 - p_2)$$

where Q is the throughput of gas to the chamber A which can be readily measured.

Hence,

$$p_1 = \frac{Q}{U}\left(\frac{1}{1-p_2/p_1}\right) \tag{3.21}$$

If p_2/p_1 is less than 0·01, $p_1 = Q/U$ and is therefore determined. Alternatively, p_2/p_1 can be considered to be known, provided the response of the gauges G_1 and G_2 is a linear function of the pressure over the range of pressures concerned. To evaluate U from the orifice dimensions, the aperture must be in a thin sheet of metal, and of area considerably less than the cross-sectional area of the chamber within which it is mounted. Otherwise, corrections must be made for an orifice in a plate of significant thickness and within a tube of comparable cross-sectional area.

Normand's apparatus (Fig. 87*b*) comprises a cylindrical test dome, of 10 inch diameter and 16 inch height, mounted above a larger chamber which is pumped by an oil vapour diffusion pump of 10 inch mouth diameter. A smaller system can readily be constructed for students' experiments on the same lines using, say, a 2 to 3 inch diffusion pump, but entailing some sacrifice of accuracy. A simple baffle is located between the pump and the orifice O. The top test chamber, A, is supplied with ports for mounting four external gauges and one nude gauge. The gas inlet line is directed upwards along the axis of the test dome, a procedure which can be shown to introduce insignificant pressure gradients along the chamber walls. Normand's orifice had a diameter of 5 cm giving a conductance of 226·7 litre per sec within the tube used, and the diffusion pump had a speed of 1,400 litre per sec. The ratio p_2/p_1 was 0·1665, insufficiently small to neglect it in equation (3.21), so linearity of the gauge over the pressure range had to be assumed. This rather large diameter orifice was needed to obtain low enough pressures in the top chamber. Smaller orifices can readily be used, but calculations on the system in conjunction with trials are necessary to ensure that the pressure range required can be covered.

To measure the gas throughput Q where the gas is dry air and the lowest pressure is 10^{-6} torr, a simple flowmeter may be used, of the type described for pumping speed measurements in section 4.1. This oil-filled flowmeter is connected to the gas inlet line to the test chamber via a needle valve. However, to calibrate an ionization gauge down to pressures of 10^{-8} torr: the test chamber of stainless steel with gauges would have to be thoroughly degassed by bake-out; the gauges degassed by eddy-current heating or electron bombardment; the oil diffusion pump provided with a liquid-nitrogen trap; metal gaskets would be preferred; and the needle valve replaced by a bakeable, metal, ultra-high vacuum valve.

Furthermore, a rather more complex gas-admittance system is

necessary to provide known, lower, gas throughputs. Here the use of a porous plug of porcelain or silicon carbide sealed within a Pyrex glass tube, as introduced by Blears (*see* Leck [103]), in the gas inlet system is useful (Fig. 88). The gas supply is best from a container of spectroscopically-pure gas. From this container, gas is admitted via a needle valve to a reservoir R of volume V, which is pumped by a small auxiliary vacuum system to a pressure p in the range 0·1 to 10 torr. As p is much larger than p_1, the pressure in the test chamber A, it equals the pressure across the porous plug of conductance U_0. The gas

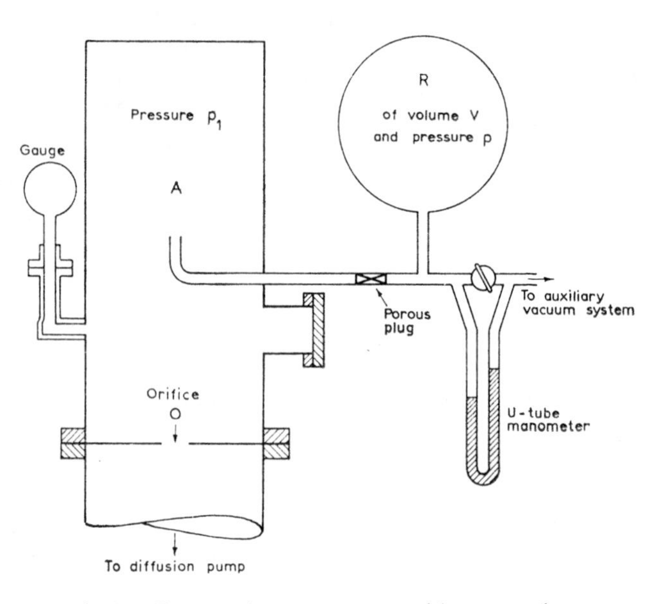

Fig. 88. Gas metering arrangement with porous plug.

throughput is hence pU_0, where p is known and U_0 has to be found. These porous materials are, in effect, a large number of capillary tubes in parallel, of diameters small compared with the m.f.p. even at pressures of 100 torr. Hence, molecular conductance prevails even at such comparatively large and easily measured pressures. Further, as the rate of flow of gas through the plug is directly proportional to the pressure p in the reservoir, it is readily shown that this pressure p at a time t is related to the pressure p_0 at the time $t=0$ by

$$p = p_0 \exp(-U_0 t/V)$$

so

$$\log_e p = \log_e p_0 - U_0 t/V$$

where V is the volume of the reservoir.

A linear graph is therefore obtained on plotting $\log p$ over the range from 1 to 50 torr, as measured by the mercury or oil manometer, against time t, for a given gas. From the slope of this straight line, U_0 for the gas concerned is found.

It is recommended, when using such a system to calibrate an ionization gauge, that the purity of the gas admitted be monitored by a mass spectrometer, and that, when the calibration is obtained, the percentages of impurities in the gas are noted.

LEAK DETECTION

The ultimate pressure p_u attainable in a vacuum chamber is decided by the effective pumping speed S_E available at p_u in relation to the quantity Q of gas entering the chamber per sec. Q will be due to actual leakage in the chamber walls or in the pumping system, due to the presence of holes, cracks, porosity, or defective seals, plus virtual leakage, due to the evolution of gas and vapour from the walls, added to which may be back-diffusion from the pump and permeability of the walls to gas. The concern is actual leakage, causing a quantity Q_L of gas per sec to enter the system, and the detection of such leaks so that they may be removed by sealing or replacement of a defective component. In general, except at ultra-high vacua below 10^{-8} torr, back-diffusion and permeability may be neglected, but virtual leakage may be considerable, and may not only adversely influence p_u but also confuse the operation of the leak-detecting method.

However, considering Q_L alone, it follows that

$$p_u = Q_L/S_E$$

where p_u is in torr, Q_L in torr litre per sec, and S_E in litre per sec; the effect of virtual leakage being ignored.

If p_u is to be 10^{-6} torr, say, and S_E is 10 litre per sec, Q_L cannot be allowed to exceed 10^{-5} torr litre per sec. Moreover, the lower the ultimate required, the smaller in proportion is the tolerable leak-rate with a given pumping speed. Alternatively, if a given leak-rate cannot be avoided, high values of S_E will be demanded to achieve a low p_u.

Leaks are undoubtedly the vacuum technician's biggest bugbear. An elusive leak may baffle him for some days. However, in the past fifteen years or so, what was once an art involving simple tests, rule-of-thumb methods, and intelligent guess-work has now become an orderly science. This technological advance has been primarily the result of utilizing, for leak detection, vacuum gauges having a selective response for a given gas, especially gas-analysing gauges such as the mass spectrometer. Furthermore, leak-detection vacuum plants designed for sensitive leak detection on components and chambers are now well established but rather expensive. If such plant is available, much time will be saved in constructing vacuum systems if all components liable to leakage are tested on it first before assembly of the system.

For practical purposes, and dealing only with the most frequently used methods valuable in the laboratory, leak-detection procedures may be divided into three groups: (*a*) simple, rough methods not involving special equipment; (*b*) methods utilizing the vacuum gauge on the system; and (*c*) sophisticated techniques involving the use of a specially built leak-detection vacuum plant for components and chambers.

3.10. *Simple Leak-detection Procedures*

(*a*) If possible, the component, e.g. a copper tube or vessel containing soldered joints, is suitably sealed temporarily, coated with soap solution, and dry compressed air is blown in at one end to indicate a hole.

(*b*) If the system is a glass one, the Tesla coil is a useful tool. The system is pumped with the backing pump alone and the Tesla coil probe is moved over the glass surface. A glow discharge is obtained, and, if the probe passes over a minute hole in the glass, a spark from the metal probe to the conducting gas will be observed.

(*c*) If the leak is probably in a glass-to-metal joint, or in metal, or for other reasons cannot be detected simply by a Tesla coil, a glow discharge is formed in a discharge tube attached to the system or in a suitable glass-walled region by means of a Tesla coil. A fine jet of carbon dioxide or methane passed over the region of the leak will then cause a change in the colour of the discharge because this gas partly replaces the residual gases in the system. Ether, alcohol, and other organic liquids may be used instead of the gas, but they tend to adhere to glass and dissolve dirt on the surface.

3.11. *Detection of Leaks by using the Vacuum Gauge attached to the System*

The **cut-off** or **isolation test** is a very useful and sensitive one, which indicates that leakage is present but does not locate the actual leaky region or hole. It can be used on almost all systems, over the whole range of pressures below atmospheric, and operates with any vacuum gauge. The pre-evacuated chamber is simply isolated from the pumps by closing a stopcock or valve and left isolated for a suitable time, after which the pressure rise is recorded. In a complex system involving several vacuum valves or stopcocks, intelligent isolation of one region at a time can lead to approximate location of the leak. A problem is that virtual leakage is also indicated. In some cases it is possible – but the operator can easily be misled – to distinguish between actual and virtual leakage, because the former gives rise to a progressive increase of pressure with time, whereas a source of vapour within the system will tend to give initially a rapid rate of pressure rise which tails off after a time to a constant level.

A sensitive gauge, such as a hot-cathode ionization gauge, in conjunction with a long cut-off time can indicate – but not locate – very small leakage. Thus, if the pressure increases from, say, 10^{-6} to 10^{-5} torr in 1 hour in a vessel of volume 5 litre, the leak rate is $5(10^{-5} - 10^{-6})/3{,}600$ torr litre per sec, i.e. $1{\cdot}4 \times 10^{-8}$ torr litre per sec.

In making such a test, the ionization gauge must be operated with an electron current less than 100 μA, to limit the gauge pumping action.

Another method, useable with any gauge, is to guess the likely location of the leak and seal it temporarily. If the guess is correct, the pressure recorded by the gauge on the pumped system will fall. Such temporary seals can be made with *Q*-compound (a plasticene-like material), silicone putty, or W-wax if heating to 50°C is permissible, or a viscous, low vapour pressure oil.

Of the various vacuum gauges described in Chapter 2, those most commonly employed for leak detection which have a sensitivity depending on the nature of the gas are the hot-cathode ionization gauge and the Pirani gauge. The cold-cathode gauge is an alternative to the hot-cathode one, but is often less satisfactory; the thermocouple gauge is an alternative to the Pirani.

A probe gas method is used in these cases. A suitable gas is played in the form of a fine jet over the walls of the system. A small container of the gas with an attached regulator valve is connected via rubber tubing to a fine jet in a drawn-down metal or glass tube. This jet of gas is passed slowly (about 1 inch in 20 sec) over the system. When it encounters a leak, the reading of the gauge increases.

The probe or search gas entering the system through the leak will change the constitution of the residual gas in the system. The sensitivity of the leak-detector element is defined (British Standards Institution 2951: 1958) as the deflection of the indicator (i.e. the meter attached to the gauge in the present case) when, at the position of the element, unit partial pressure of air is exchanged for unit partial pressure of the probe gas. This sensitivity will increase the more completely the leak is covered by the probe gas; the higher the sensitivity of the gauge to this gas; the lower the gas viscosity, so that it enters the leak readily; and the smaller the pumping speed of the system for this gas.

With a hot-cathode ionization gauge, butane (Calor gas) meets this requirement best, though propane, helium, hydrogen, and trichlorethylene vapour from a piece of soaked cotton-wool around a rod of glass may all be used. Using a Bayard-Alpert gauge with an electron current below 100 μA and butane probing, it is possible to locate leaks in a system pumped to between 10^{-4} and 5×10^{-9} torr. Careful use of this method will often detect a leak which prevents a chamber from being pumped to below 5×10^{-9} torr by means of a pump of effective speed 5 litre per sec. The ability to find such small leaks will depend, however, on the gas flow in the system, and so on the relative positions of the gauge, the inlet to the pump, and the site of the leak. To obtain increased ease of location, the pumping speed may be reduced by partially closing the isolation valve between the chamber and the pumps, and it is recommended that the steady ionization current at the prevailing pressure as recorded by the d.c. amplifier be backed-off, because it is clearly easier to detect a change of this current from near zero than from a finite value.

With the Pirani gauge, hydrogen with its low viscosity and high thermal conductivity is the best probe gas. The Pirani gauge is useful at pressures in the range from 10 down to 10^{-3} torr in a system pumped by an oil-sealed rotary pump alone, where partial throttling of this pump by a valve is helpful. It can also be used advantageously in the backing stage of a vapour pump/rotary backing pump combination, in which case the pressure above the vapour pump may be in the range from 10^{-3} to 10^{-7} torr. The best sensitivity is obtained with pressures at the Pirani gauge head of below 10^{-1} torr. To use this procedure, the Pirani gauge is best balanced at the pressure prevailing – which should be as steady as possible – and a more sensitive instrument such as a lamp-and-scale galvanometer substituted for the normal panel-mounted microammeter. Butane is better than hydrogen if the speed of the pump can be artificially restricted, but hydrogen is normally used. Coal-gas is a simply-obtained alternative, but gives about one-tenth the sensitivity of hydrogen.

With the Bayard-Alpert gauge and butane probing, leaks as small as 10^{-7} torr litre per sec and below can be detected; the Pirani gauge with hydrogen probing is able to detect leaks of about 10^{-5} torr litre per sec.

The smallest leak that a leak detector can indicate is quoted in terms of the minimum throughput of air into the leak that is detectable. This may be expressed in various units depending on the unit of pressure and the unit of volume chosen. Those most commonly employed are the torr litre per sec, the lusec, and the standard cu cm per sec. The word 'lusec' comes from litre micron per sec, i.e 1 μ sec. Note:

$$\begin{aligned} 1 \text{ standard cu cm per sec} &= 1 \text{ cu cm of gas at 760 torr per sec} \\ &= 0{\cdot}76 \text{ torr litre per sec} \\ 1 \text{ lusec} &= 10^{-3} \text{ torr litre per sec} \end{aligned}$$

If a system is pumped by a cold-cathode getter-ion pump, the current passing through the gaseous discharge in the pump is a linear function of the pressure in the pump; no additional vacuum gauge is necessary, presuming the pressure in the pump is the same as that in the chamber. The meter, in series with the H.T. supply to the pump electrodes, which records the pump current may therefore be calibrated so that it also records the pressure in the pump. Furthermore, the current I through such a pump at a given pressure p when operating against an air leak in the system will change significantly by ΔI when another gas – the probe gas – is admitted through this leak. If the probe gas is argon, nitrogen, or helium, the current increases; if it is oxygen or carbon dioxide, this pump current decreases.

To be able to record small values of the change ΔI and thus small leaks, the pump current before probing must be constant. Therefore, the initial gas conditions in the system and the H.T. supply to the pump must be stable. Furthermore, a stabilized amplifier is needed to amplify and record small values of ΔI.

This technique of leak detection was originated by Varian Associates Ltd. and is the principle of their **VacIon pump leak detector.** This unit contains a stabilized H.T. supply and an electronic amplifier. For this firm's 5 litre per sec and 8 litre per sec VacIon pumps (Vac Ion is their trade name for a cold-cathode getter-ion pump), the leak-detector unit can supply the H.T. and the indication of current change ΔI on gas probing for leaks. Larger pumps demanding high initial pump currents are started with their appropriate H.T. units and then the leak-detector unit substituted, of which the H.T. power-pack supplies 3,300 volt on open-circuit and a short-circuit current of 150 mA.

Maximum sensitivity in leak detection is obtained by using a potentiometer recorder at the output of the amplifier, and the sequential use of first oxygen (ΔI negative) and then helium (ΔI positive). Leaks may be found over the pressure range from 10^{-4} to 10^{-11} torr and with a reported maximum sensitivity of $2{\cdot}6 \times 10^{-11}$ standard cu cm per sec of air, which is 2×10^{-11} torr litre per sec.

3.12. *Special Devices and Leak Detection Vacuum Systems*

Of the several devices which have been developed for leak detection, those most widely used, other than standard vacuum gauges, are the palladium barrier leak detector, the halogen leak detector and the mass spectrometer. If convenient, any one of these devices may be connected to a vacuum system for leak-detection purposes. More sophisticated practice, however, is to use the device within a vacuum system specially constructed for leak detection on components or chambers; the system is then designed to obtain maximum sensitivity of detection and to avoid spurious effects that may adversely affect optimum operation.

The **palladium barrier leak detector** (Nelson [107]) was first made commercially available by the Radio Corporation of America and has been developed in Britain by Edwards High Vacuum Ltd., whose gauge head is shown in outline in Fig. 89. This head is of glass with a seal to Kovar for joining it to a system via an O-ring. The end of this Kovar tube within the glass envelope is sealed by a thin disk of palladium. The glass envelope contains the electrodes of a hot-cathode ionization gauge of special construction, is permanently sealed-off under vacuum and is gettered. The electrons from the hot thermionic cathode bombard the palladium disk, which is an anode at about +400 volt with respect to the cathode, and raise the temperature of this disk to about 800°C, with an anode dissipation of 6 watt. The palladium at this temperature becomes permeable to hydrogen but not to other gases. Around the axially mounted cathode and anode is a metal cylindrical collector maintained at between 0 and −90 volt relative to the cathode. This forms a collector for positive ions and also assists by focusing the electrons at the anode. If this gauge head is attached to a system containing a leak, and a fine jet of hydrogen as the probe gas enters this leak, some of it will permeate the palladium anode, and current will be recorded at the ion collector due to the positive hydrogen ions formed and collected. The marked increase of this ion-collector current above that due to the residual pressure of about 10^{-7} torr in the gauge head indicates the presence of the leak.

As the low pressure in the gauge head is independent of the pressure in the system, the latter can be much higher than the former; indeed, the system can be pumped by an oil-sealed rotary pump alone. However, the chief difficulty is then that hydrocarbons from the pump oil and water vapour are dissociated at the hot palladium to produce

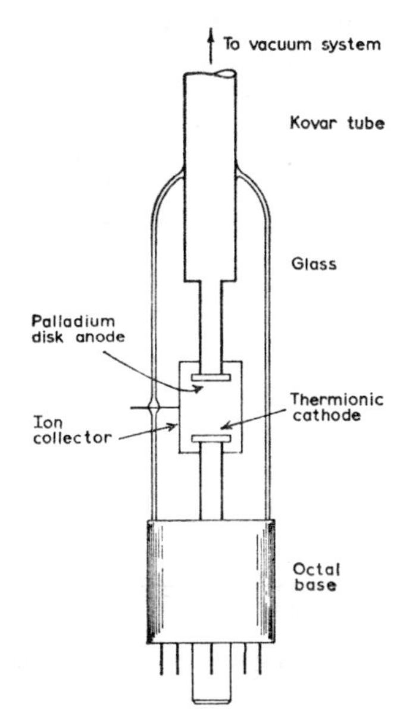

Fig. 89. Palladium barrier gauge head.

hydrogen, which is confused with the hydrogen entering the leak on probing. This problem is overcome by installing a cold trap cooled with liquid nitrogen or solid carbon dioxide in the tubulation between the rotary pump and the gauge head. A useful leak-detecting arrangement is then available.

The palladium barrier gauge is capable of detecting leaks as small as 10^{-8} torr litre per sec. To obtain this sensitivity, however, two additional factors have to be considered in designing a leak-detection vacuum system. The first is that the oil-sealed rotary pump suffers small pumping speed fluctuations as its rotor revolves: such fluctuations hinder the detection of small leaks. It is preferable, therefore, to connect the sensing gauge head to the intake of a vapour pump which is then backed by a rotary pump. The second factor is that increased sensitivity is obtainable if the gas from the chamber is compressed before reaching the gauge head. This compression or pressure amplification is provided by a vapour pump. The use of the palladium barrier gauge in a sensitive leak-detection plant is therefore best with the gauge head connected via a liquid-nitrogen-cooled cold trap to a ballast volume (which inhibits pressure fluctuations) between *two* oil vapour diffusion pumps. The first vapour pump D_1 provides the pressure amplification with a consequent increase of sensitivity of

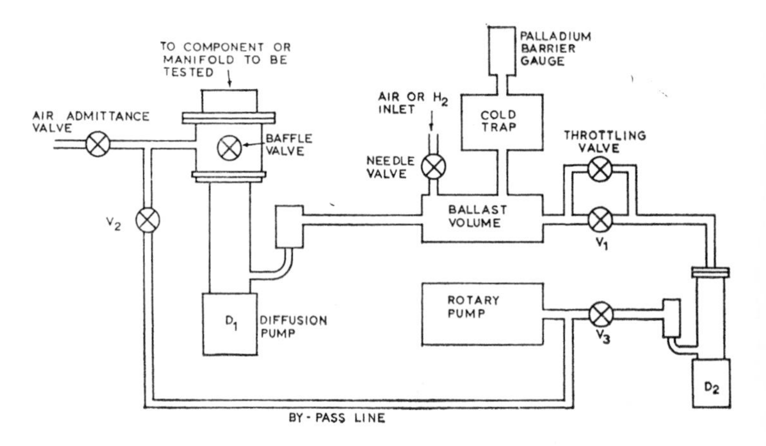

Fig. 90. Leak-detection vacuum system utilizing a palladium barrier gauge.

about × 30, the second D_2 smooths out the pressure fluctuations of the rotary backing pump. In a system of this kind described by Ochert and Steckelmacher [108] (Fig. 90), there is also provision of a needle valve to the ballast volume to which the gauge head is connected. This enables hydrogen to be trickled into the system to add at will to the amount of hydrogen present resulting from probing the leak. The necessity for such additional hydrogen arises because oxygen sorbed on the palladium may restrict the permeation of hydrogen.

The operation of the **halogen leak detector** (White and Hickey [109]) depends on the fact that platinum (as also does glass, which can be used) emits positive ions when heated *in vacuo*, and this positive ion emission is much affected by the presence of halogens.

To detect leaks in vacuum systems, a leak detector tube is used containing a platinum cylindrical *anode* which is heated indirectly by an inside filament carrying electric current. The positive ions emitted are collected by a surrounding coaxial cathode (Fig. 91) at a negative potential with respect to the anode of 50 to 500 volt, with a d.c. amplifier or microammeter in series to register the positive ion current. The probing gas is then a halogen-bearing one such as freon 12, carbon tetrachloride, or trichlorethylene. When this probe gas – freon 12 is preferred – enters a leak, the positive ion current increases, so the presence and location of the leak is recorded.

The halogen method is valuable in that the positive ion emission from platinum is independent of the surrounding pressure provided the gas is halogen free. The detector tube can therefore be at any pressure from atmospheric down to the ultra-high vacuum range.

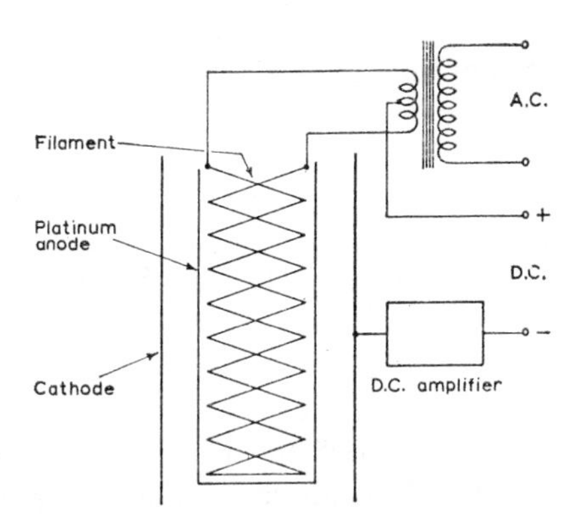

Fig. 91. A halogen leak-detector tube.

However, if operated at, say, 1 torr, in a large container, leak detection will be made difficult because of the long time elapsing between entry of the halogen through the leak and its effect on the positive ion emission.

In a leak-detection vacuum plant utilizing a vapour pump/rotary backing pump combination, the halogen detector tube may be connected to either the high vacuum side – giving a shorter response time – or the backing vacuum side, providing higher sensitivity because of pressure amplification.

This technique can also be used to find leaks in containers which can be filled with air and freon at a pressure exceeding atmospheric.

The halogen leak-detector electrodes are then mounted in a container at atmospheric pressure furnished with a handle to look like a pistol. A narrow steel tube enters this container, and the pistol 'sniffer' is moved over the container walls until a leak is detected, when freon escapes through a hole to enter the narrow tube connected to the interior electrodes.

The minimum detectable leak by the halogen detector tube operated on a vacuum system is about $2{\cdot}5 \times 10^{-8}$ torr litre per sec. It is an alternative to the palladium barrier method, preferred by some workers because it is not so suspect to spurious effects arising from hydrogen generated within the system.

The most sensitive vacuum systems for leak detection in components are based on the **mass spectrometer**. The usual procedure is to use a magnetic-deflection-type mass spectrometer 'tuned' to record a mass of 4 a.m.u., i.e. helium, and then employ helium as the probe gas. This method is not only of great sensitivity, it is also the most reliable means of indicating small leaks because helium is unlikely to be present in the system for any other reason than entry of the probe gas.

Several vacuum manufacturers market mass spectrometer leak-detection systems; most of them are based on a sector field type of magnetic deflection instrument. The 180° deflection instrument specially designed in a small compact assembly with a permanent magnet is popular. An example is the 'Centronic' mass spectrometer leak detector of 20th Century Electronics Ltd., in which the radius of the semicircular ion path is 1·25 cm. This is similar to the Consolidated Electrodynamics Corp. detector, which employs a Diatron miniature 180° mass spectrometer.

The head of the Centronic mass spectrometer (Fig. 92) contains a thermionic filament F, from which electrons are accelerated through slits in the positive box B to be collected at an anode A. Positive ions formed in the wake of these electrons are repelled by the positive repeller plate R and attracted to the slit S in the plate N. This slit S is parallel to the electron beam between F and A. The accelerated positive ions which traverse the slit S then execute (at constant velocity for a particular mass-to-charge ratio) a semicircular path in electric field free space within the perpendicular uniform magnetic field, of flux density 2,200 gauss approximately, provided by a surrounding permanent magnet. When the accelerating p.d. on the repeller plate is about 125 volt, singly-charged helium ions of mass 4 a.m.u. describe a semicircle of such a radius that they are focused at the exit slit Z, which they traverse to reach the ion collector plate C after passing another slit in a suppressor plate D. These ions form a current which passes down an external resistor of 10^{12} ohm to earth. The p.d. developed is the input to a d.c. amplifier with an electrometer valve input stage. The output meter then suffers a deflection which is proportional to the partial pressure of helium in the system. Only singly-charged helium ions will be accurately guided towards the collector plate C; other ions, of different mass-to-charge ratio, will execute paths in the magnetic field of the wrong radius and so will not traverse the exit slit Z. However, scattered ions of other than helium

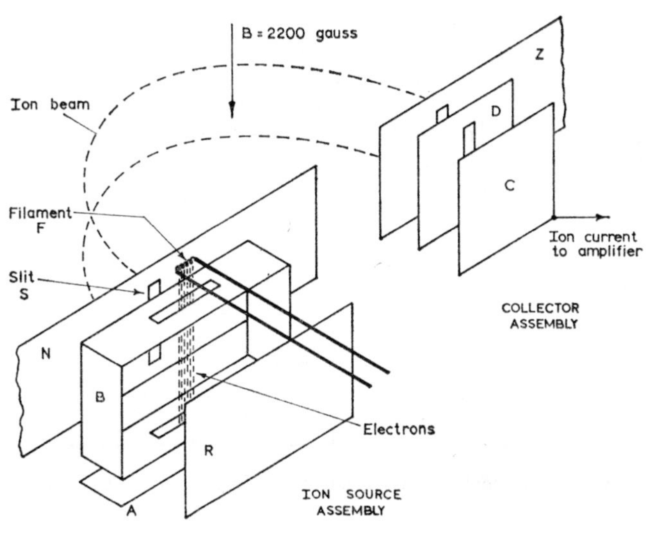

Fig. 92. 'Centronic' mass spectrometer head.

may traverse Z by chance. These can be prevented from reaching C by a positive potential of about 20 volt on the suppressor D, because they have small velocities compared with the helium ions of energy 125 eV.

This type of mass spectrometer cannot be operated at a pressure exceeding 10^{-4} torr because of the thermionic filament, but it is valuable at pressures down to 10^{-10} torr and below. The head is provided with its own vacuum system, quite apart from that of the system used to evacuate the chamber or component under test. A general arrangement of this **mass spectrometer leak detector vacuum**

system (Fig. 93) comprises a liquid-nitrogen-filled cold trap, an oil diffusion pump and a rotary backing pump with isolation valves and a Bayard-Alpert gauge. This system is best of polished stainless steel to avoid corrosion and outgassing. The component or chamber under leak-test is evacuated by a separate vacuum system where a single oil-filled rotary pump is often adequate. The usual procedure is to connect this component or chamber to the vacuum system of the mass spectrometer by tubulation which contains a needle or throttle valve. The helium probe gas-jet is then moved slowly over the component or chamber. When this gas-jet encounters a leak, a pressure rise is re-

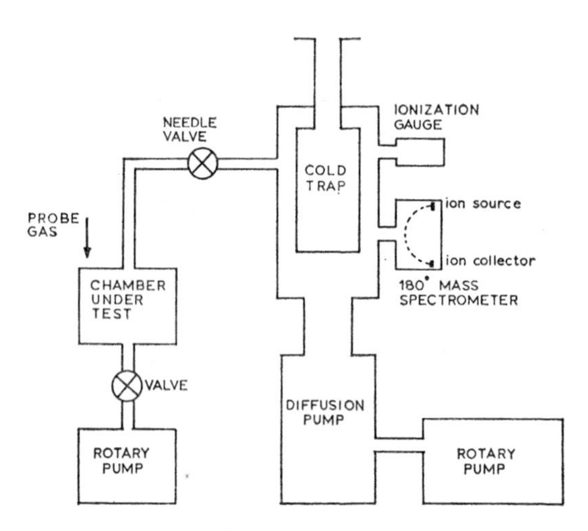

Fig. 93. Arrangement of mass spectrometer leak-detector vacuum system.

corded by the output meter or recorder of the mass spectrometer d.c. amplifier.

Alternative procedures follow.

(*a*) In some cases, the vacuum system of the mass spectrometer can also be used to evacuate the component or chamber under test. This is only possible if this component is not too large and has been first tested by other means to ensure that it is free of large leaks, otherwise the pressure in the mass spectrometer will rise above 10^{-4} torr.

(*b*) A 'sniffer' technique may be used if the component or chamber under test can be pressurized with helium or gas containing helium. A sampling probe is then connected by flexible tubing to the inlet of the mass spectrometer system. The sampling probe is then moved over the outside surface of the component; when it encounters a leak, helium enters the mass spectrometer.

The mass spectrometer method is capable of detecting leaks as small as 10^{-10} to 10^{-11} torr litre per sec, and much smaller with more elaborate instrumentation.

To leak-test a number of components or chambers, a manifold or base-plate arrangement is convenient and may be adapted to any one of the leak-detection vacuum systems. For example, a manifold with a number of port-holes may be set up and the components sealed over these holes by suitable gaskets or sealing compounds. Again, a base plate with sealing gaskets of various sizes may be used as an accessory to leak-test a number of vacuum chambers of the bell-jar type. The **hood test** is also useful: here a cup or bell-jar or even a plastic bag is placed over the component or chamber which is connected to the leak-detection system. The probe gas is then admitted inside this hood. This enables a test to be made of whether the component is leaky or not, but does not actually locate the leak. Subsequently, the leak can be located by the probing method; often this is unnecessary because the component is simply rejected as not vacuum tight. The plastic bag method is convenient in testing a large chamber: however, if hydrogen is used as the probe gas, it is not recommended because of the danger of sparks.

Audible alarm systems can readily be devised for inclusion within leak detectors having an electrical output. The current change on encountering a leak with the probe gas is amplified and fed to a small loudspeaker. This provision is useful in that the technician may well be probing for a leak in a location from which he is unable to see deflection of the output meter easily.

CHAPTER FOUR

THE MEASUREMENT OF PUMP PERFORMANCE

4.1. *The Measurement of Pump Speed*

Adopting a more specific definition than that given on p. 6, **the speed S of a pump for a given gas** is defined as the ratio of the throughput Q of that gas to the equilibrium pressure p at the mouth of the pump under specified conditions of operation. Thus, $S=Q/p$. This definition applies to any type of vacuum pump. For mechanical rotary pumps, S is quoted in litre per min and Q in torr litre per min. In the case of vapour pumps and other types providing pressures below 10^{-3} torr, S is quoted in litre per sec and Q in torr litre per sec.

The speed of a pump varies with the pressure prevailing at its intake port, though the variation is often not significant over some part of the pressure range. It is therefore necessary to obtain a speed-pressure characteristic. To determine this, the practice is usually to set the gas pressure p at various equilibrium values within the working range, by metering gas into a test header connected to the pump inlet, and achieve the equilibrium by balancing this inlet throughput Q of gas against the speed S of the pump.

To measure the equilibrium pressure, two types of gauge are normally employed: the McLeod gauge, for the range from 10 to 10^{-4} torr; and the hot-cathode ionization gauge, for pressures below 10^{-4} torr. If it is required to record pressures above 10 torr, a mercury or oil manometer is used.

The McLeod gauge must be trapped, i.e. a cold trap chilled with liquid nitrogen or air needs to be installed in the connecting tube between the test header and the gauge. Mercury vapour from the McLeod gauge is then excluded from the system and this gauge will record only permanent gas pressures.

The hot-cathode ionization gauge must be degassed by electron bombardment and run with an electron current of 0·1 mA or less, to avoid significant electrical pumping by this gauge. Further, it is preferable to use a Bayard-Alpert-type gauge with a low temperature emitter filament, e.g. coated with lanthanum hexaboride, to minimize chemical reactions between the gas and the filament. As sorption of oxygen is significant at hot-cathodes, it is also recommended that any tubulation between the gauge head and the test header should have a conductance exceeding 50 litre per sec, to avoid pressure difference between the gauge electrodes and the header.

The necessary throughput Q of gas into the test header has to be determined. This is done by a flowmeter of which the displacement type is most used. Usually the gas admitted is air (dried by passing

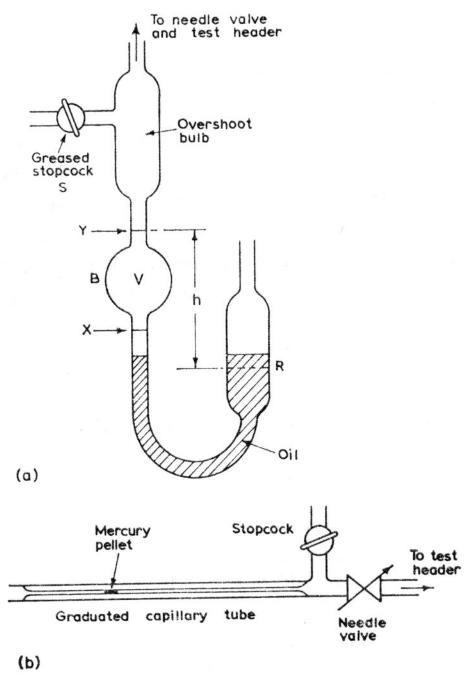

Fig. 94. Displacement type gas flowmeters.

over silica-gel), but other gases from a suitable reservoir can be employed.

A type of flowmeter often used is that shown in Fig. 94(*a*). This is made of glass and filled with oil of the same type as used in the pump (silicone oil serves with a mercury diffusion pump). The left-hand limb of the flowmeter contains a bulb B of known volume V between lower and upper graduation marks X and Y respectively. At atmospheric pressure, the oil level is below X and equal to that in the reservoir R. Above the bulb B is an 'over-shoot' bulb to which is attached a greased stopcock S by means of a side-tube. Above the over-shoot bulb, the flowmeter is connected by rubber pressure tubing to a needle valve which connects directly into the inlet tube to the test header.

Irrespective of the type of pump, stopcock S is initially open and the needle valve closed. The pump is then switched on and an equilibrium ultimate pressure and temperature established. The needle valve is then opened partially to establish a set pressure p in the header. When this new equilibrium pressure is attained, the stopcock S is closed, then the oil in the flowmeter will rise into the bulb B. The time t for the oil level to rise from mark X to mark Y is measured with a stopwatch, where the volume V between X and Y is known. After ensuring that t has been measured satisfactorily, the stopcock S is opened to restore equality of oil levels in the two limbs of the flowmeter, and the needle valve is re-adjusted to give a new pressure p in the header. The readings are then repeated.

The air in the flowmeter between the lower mark X and the needle valve is initially at atmospheric pressure p_a torr (recorded by a Fortin barometer) and has a volume V_1. Finally, it is at a pressure $(p_a - h\rho/13{\cdot}6)$ torr and has a smaller volume V_2, where h is the head of oil of density ρ between the upper mark Y and the now reduced level in the reservoir. The throughput Q of gas is therefore given by

$$Q = \left[p_a V_1 - \left(p_a - \frac{h\rho}{13{\cdot}6}\right) V_2\right] \bigg/ t$$

As

$$V_1 - V_2 = V$$

$$\therefore\ Q = \left[p_a V + \frac{h\rho}{13{\cdot}6}(V_1 - V)\right] \bigg/ t \qquad (4.1)$$

$h\rho/13{\cdot}6$ is small compared with p_a because ρ is about unity, h about 100 mm, and p_a is nominally 760 torr. Further, $(V_1 - V)$ can be made small. Hence, equation (4.1) reduces with small error to

$$Q = p_a V/t$$

This must equal the throughput into the test header at the much lower pressure p; the pressure drop from near atmospheric to the vacuum in the header being achieved by the needle valve. The speed S of the pump is then given by Q/p, i.e.

$$S = p_a V/pt$$

where S is in litre per sec, provided that p_a and p are in the same units (usually torr), V is in litre, and t in sec.

A graph of S against p can therefore be obtained for the pump in question and is usually plotted on log-linear graph paper with the horizontal logarithmic axis marked in torr and the vertical linear axis in litre per sec.

With a flowmeter of the type shown, the time t can only be recorded with accuracy if it exceeds 5 sec and is preferably not greater than, say, 10,000 sec. Assume p_a is 760 torr, then with a convenient value of V of 0·01 litre the speed range in litre per sec coverable by the flowmeter is from $7{\cdot}6\times10^{-4}/p$ to $7{\cdot}6/5p$. At $p=10^{-5}$ torr, this corresponds to a range from 76 to 152,000 litre per sec, covering all vapour pumps except the smaller ones. At $p=10^{-4}$ torr, the range is 7·6 to 15,200 litre per sec. As the pressure is increased, so these figures at the extreme of the range decrease in proportion. It is seen that for the measurement of the speed of the smaller pumps at low pressures, V must be reduced to 0·001 litre or less. 1 ml can be achieved by choosing marks X and Y on a straight tube in place of the bulb C. Alternative practice for low speeds at low pressures is to use the simple flowmeter of Fig. 94(*b*), which takes the form of a graduated capillary tube in which is placed a pellet of mercury, or preferably, a close-fitting, small, steel ball lubricated with oil (A. Leemans: private communication), which can be conveniently moved to any position along the capillary by means of a magnet.

As the set pressure p is increased to above 10^{-1} torr, the speed measurable with a displacement flowmeter and with convenient times of oil level rise becomes restricted. Indeed, for gas throughputs exceeding about 0·1 torr litre per sec, the displacement flowmeter is inadequate as the rise times become too short. A commercial flowmeter of the orifice or float type is then used.

The **design of the test header** is important in that it is vital to avoid any direct line between the gas inlet to the header and the connection to the gauge, because the admitted gas may then be partly beamed into the gauge and give false pressure readings.

The test header used when **measuring the speed of an oil-sealed mechanical rotary pump** is conveniently of circular cross-section. It is usually chosen to have a volume a few times greater than the volume swept by the rotary pump in one compression cycle, so as to minimize pressure fluctuations due to the pulsing effect consequent upon revolution of the rotor. To avoid any direct line between the gas inlet and the gauge connection, the gas inlet from the external needle valve

N is conveyed into the header via an internal axial tube containing a right-angle bend and leading upwards to a funnel-shaped outlet near the top cover of the test header. The dimensions and positioning are shown in Fig. 95. Further, the connection to the gauge (usually a McLeod) is from an internal tube containing a right-angle within the header and terminating at an inlet on the central axis below the bend in the gas-flow internal tube. In this way, incoming gas molecules must be scattered from the header top before affecting the pressure in the vicinity of the inlet to the gauge tube, and the arrangement simu-

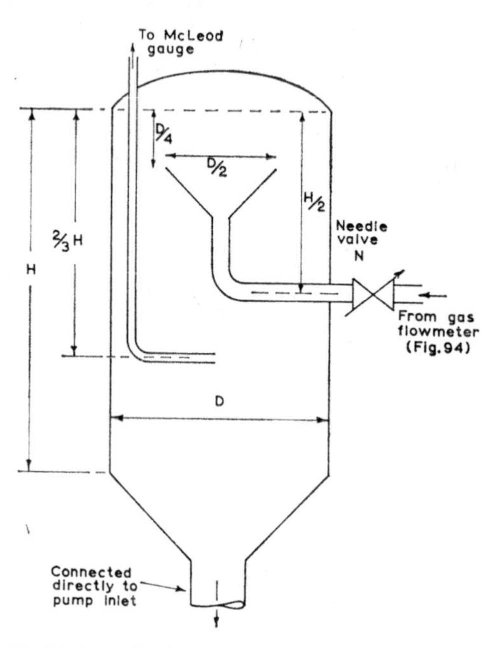

Fig. 95. Test header for use on a mechanical rotary pump.

lates a chamber in which the gas molecules have velocity vectors which are completely random in distribution.

The test header used when **measuring the speed of a vapour pump** is of great importance and has been the subject of much controversy. Ideally, all the gas molecules which enter the pump mouth should do so with no preponderant flow. This means that the molecules should have a wholly random distribution over all directions, so that the number of molecules leaving a surface in a direction making an angle θ with the normal to the surface is proportional to $\cos \theta$. When the gas is confined within a test header of finite volume, there will arise some streaming effect of the molecules into the pump mouth. The ideal, whereby a Knudsen cosine distribution of molecules is obtained at the pump mouth, can be realised by using a large test header with suitably positioned gas inlet and gauge. In practical vacuum systems such conditions often do not obtain; again, in the specification of a test header, provided that it gives results which do not depart too greatly from the ideal, it is more important to arrive at an agreed arrangement rather than insist on the use of chambers of such dimen-

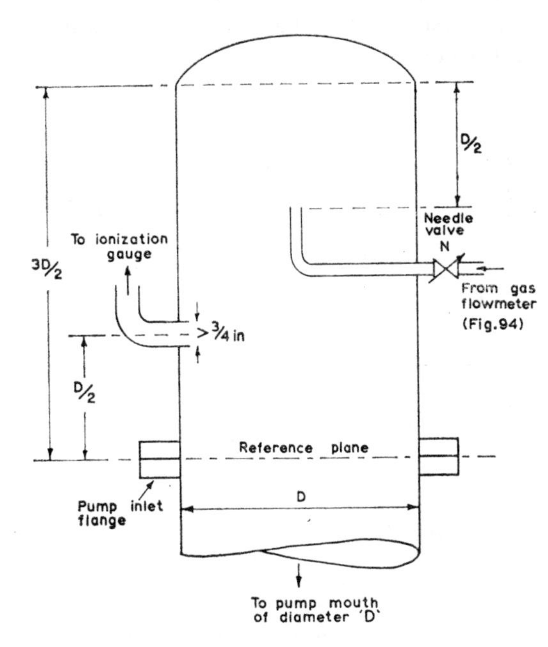

Fig. 96. Arrangement for determining the speed against pressure characteristic of a vapour diffusion pump.

sions as to be difficult and costly to fabricate. Agreement is necessary: two vapour diffusion pumps of the same mouth diameter and physical size may be specified to have considerably different speeds; it is possible that the disparity is due to different measurement techniques adopted and not that one design is more efficient than the other.

The present tendency is to use a test header conveniently of the same diameter as that of the pump mouth, and to position the gas inlet and gauge positions so as to simulate closely the conditions that would exist in an ideal chamber. Fig. 96 shows such a test-header: this is a cylinder of the same diameter D as the pump mouth and of height $\frac{3}{2}D$. The connecting tube to the gauge enters the vertical wall of the header along its radius and with its axis at a height $\frac{1}{2}D$ above the pump inlet flange. This tube should have a length not exceeding 6 inch and a diameter not less than $\frac{3}{4}$ inch. The gas inlet arrangement is such that molecules of gas are scattered from the top of the header with a distribution of velocity vectors following as closely as possible the Knudsen cosine law. This is achieved by a pipe entering the header along its radius and bent upwards through a right-angle to lie along the header axis and terminated at an outlet a distance of $D/2$ from the flat-top cover plate.

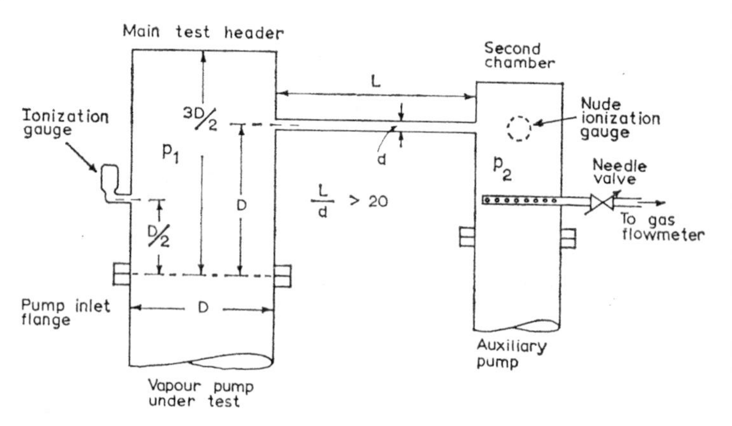

Fig. 97. Speed testing by a conductance tube method.

For small flow rates of less than about 5×10^{-4} torr litre per sec, the method described above leads to inconveniently long times of measurement. A variation on the procedure, which is preferable then, is to admit the gas from a reservoir at a known low pressure through a tube of known conductance into the test header. This set-up is only useful for measuring speeds at pressures below 10^{-3} torr, and has been found valuable in speed studies on getter-ion pumps as well as vapour diffusion pumps.

Speed testing by the conductance tube method is carried out by an arrangement shown in Fig. 97. The test header on the pump and the gauge position are the same as in Fig. 96, but the gas inlet tube now enters the side of the header at a height D above the pump flange. This inlet tube is cylindrical and of known diameter d and length L;

where L/d is preferably greater than 20. Its inlet end is from a second chamber above an auxiliary pump. A nude ionization gauge enters this second chamber as shown, whilst the gas inlet from a needle valve outside this second chamber is into a horizontal internal tube containing a series of holes. In this way, a random distribution of gas molecules is ensured in the second chamber.

Let p_1 equal the equilibrium pressure measured in the main test header, when the gas throughput Q is established due to a pressure $p_2\,(>p_1)$ in the second auxiliary chamber brought about by a suitable gas inlet rate through the needle valve. If p_{o_1} is the ultimate pressure in the main test header and p_{o_2} that in the auxiliary chamber, the flow rate Q into the main test header is given by

$$Q = U[(p_2 - p_{o_2}) - (p_1 - p_{o_1})]$$

The tube conductance U is calculated knowing its length L and diameter d by means of equation (3.16). This conductance must be small compared with that of the main test header and the pump speed. Further, if U is chosen so that $(p_2 - p_{o_2})/(p_1 - p_{o_1})$ is not less than 100, the previous equation simplifies to give

$$Q = U(p_2 - p_{o_2})$$

The speed S of the pump under test is given by

$$S = \frac{Q}{p_1 - p_{o_1}} = \frac{U(p_2 - p_{o_2})}{p_1 - p_{o_1}}$$

Hence S is found at the test header pressure p_1, because U is calculated, and p_2, p_{o_2}, p_{o_1} are all measured. Note that the ultimate pressures p_{o_1} and p_{o_2} are important with low gas throughputs, obtained when the pump is working near its ultimate pressure. In some cases, however, p_{o_1} and p_{o_2} may be neglected as less than 2% of p_1 and p_2 respectively. It must be stressed that p_1 and p_2 must both be molecular pressures ($< 10^{-3}$ torr) at which the m.f.p. exceeds the tube diameter d, otherwise the molecular conductance equation for U does not apply.

4.2. *The Measurement of Ultimate Pressure*

For an oil-sealed rotary pump, the ultimate pressure usually specified is that due to the permanent gases. As the oil in the pump will exert a vapour pressure, it is best to incorporate a liquid-nitrogen trap in the line between the gauge and the pump intake port. If this trap is not included, it is essential to check the compression ratio of the McLeod gauge to find out whether or not any vapour present is compressed to saturation: if it is, the saturated vapour pressure may be an appreciable contribution to the pressure in the compressed gas; if not, the vapour pressure may be recorded. The McLeod gauge is suitable for pressures in the range 10 to 10^{-4} torr; for ultimates below 10^{-4} torr, a degassed hot-cathode ionization gauge (with trap) is used.

The McLeod gauge must be joined via the cold trap (the U-tube pattern of Fig. 22(*a*) is suitable) to the pump (furnished with an isolation valve) by means of leak-free tubing. Rubber pressure tubing is not recommended; flexible metal bellows tubing or glass tubing with suitable, greased cone-joints is preferred.

If the pump is fitted with gas-ballast, it should be run for 8 hour or more with the ballast full on and the isolation valve closed to free the pump oil of contaminants.

To record the ultimate pressure, the gas-ballast valve is closed, the isolation valve opened and the cold trap chilled with liquid nitrogen. To enable the McLeod gauge to be kept under vacuum when not in use, a greased stopcock is included above it. When the pressure produced by the pump is below about 1 torr (attained in a few min), the stopcock to the McLeod is opened. The pressure is then recorded at hourly intervals. The ultimate pressure attained by the pump is that recorded when three successive pressure readings indicate no further reduction in pressure. It must be ensured that the system is free of leaks. The measurement of the ultimate pressure is repeated with the gas-ballast full on.

It is generally misleading to quote an ultimate pressure for an oil vapour diffusion pump, because this will depend on the design of the system within which it is incorporated. Thus, a pump with a nominal ultimate pressure of 10^{-6} torr can be used to achieve ultra-high vacua if it is effectively baffled and trapped, suitable sealing gaskets are used, and the system is bakeable (section 3.8). Further, a mercury diffusion pump will not provide an ultimate pressure below the vapour pressure of mercury at the ambient temperature (10^{-3} torr at 15°C) unless it is used with a cold trap.

The hot-cathode ionization gauge is usually employed to measure ultimate pressures of vapour pumps. The Bayard-Alpert gauge is preferred and must be thoroughly degassed. This gauge must be connected to the pump test header in such a way that there is no direct line between the pump mouth and the electrodes. Further, the conductance for air of any tubulation between the gauge electrodes and the pump mouth should exceed 50 litre per sec, otherwise significant differences of pressure can become established between the gauge and the pump because of the pumping action of the gauge. The electrical pumping action due to positive ions in the gauge becoming retained as gas at the ion collector and its walls can be reduced to insignificant values by limiting the electron current to below 0·1 mA. However, the gauge hot filament can sorb active gases, particularly oxygen and hydrogen, so the oxygen in the air may be pumped rapidly. Indeed, there is a growing, healthy tendency to suspect the usual hot tungsten filament of an ionization gauge, owing to its activity in the presence of oxygen, carbon dioxide, water vapour, and hydrogen, which becomes partly atomic hydrogen with a high affinity for wall surfaces. The present remedy is to use a lanthanum-hexaboride-coated filament run at a temperature of about 1,000°C or less to minimize these reactions (see also p. 129).

4.3. *Measurement of the Back-streaming Rate of an Oil Vapour Pump*

Among the several possible measurements of the various aspects of pump performance, there is space for only one other important quantity: the rate at which oil back-streams from an oil vapour pump. A method described by Power and Crawley [110] utilizes a cylindrical test header of diameter equal to that of the pump mouth and with a coned-top cover plate (Fig. 98). Between the flange to this header and the flange of the vapour pump is inserted a removable flange furnished with a machined, annular, collecting channel. Sealing between the three flanges is by O-rings, as shown. The oil which back-streams into this header collects in this channel, and then drains into a graduated burette. The other end of this burette connects to a valve. From this valve, there is a tube leading to backing connection of the vapour pump, so that oil collected can be drained back into the vapour pump on opening the valve.

In the test procedure, the system is run for 48 hour or more to ensure that the inside of the header is coated with a thin film of the pump oil. The test proper is then begun and oil is collected over a period long enough (usually 70 hour or more) to ensure sufficient quantities for accurate measurement. The test header should be maintained at the same temperature as the pump cooling water at the inlet, or at room temperature for an air-cooled pump. It is advantageous to tilt the pump and header axis slightly from the vertical, to facilitate drainage of the oil into the burette. The back-streaming rate is quoted usually for a given oil in ml per sec per sq cm of the pump mouth cross-sectional area.

It is also important to be able to measure the reduction of back-streaming brought about by the use of a baffle above the oil vapour pump (section 1.11). The use of an efficient baffle will reduce the

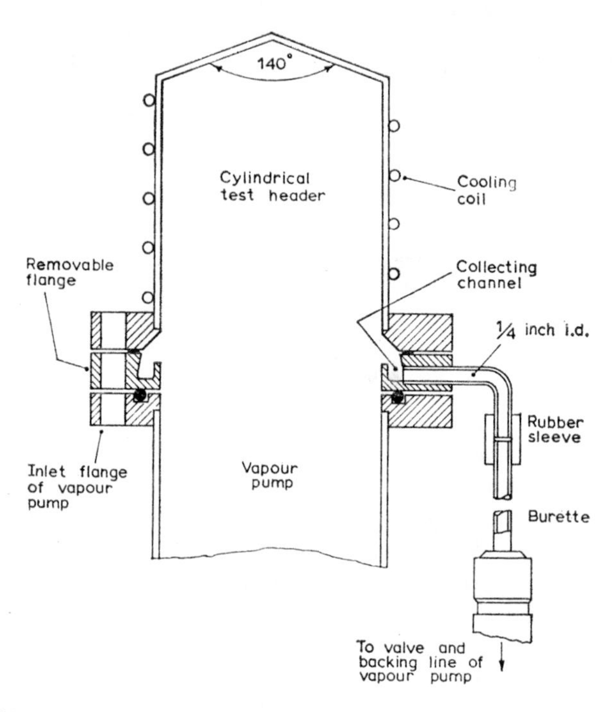

Fig. 98. Measurement of the back-streaming rate of an oil vapour pump.

quantity of back-streaming oil to such low values that a direct gravimetric method is insufficiently sensitive. In recent work, the measurement of much smaller amounts of back-streaming oil has been undertaken, particularly by the use of a quartz crystal in the pump header. The crystal sensitively undergoes a change of its high-frequency resonant vibration on accumulation on its surface of very small amounts of oil.

CHAPTER FIVE

MISCELLANEOUS MATERIALS, DEVICES, AND PROCESSES

Within the area of vacuum technology and its applications, there is encountered a very wide variety of materials about which it is important to have data, particularly as regards their properties and behaviour within an environment at low gas pressure. These materials may require special processes for handling and cleaning. Again, certain devices of great assistance in various aspects of vacuum technique merit description and do not necessarily come within the province of the production and measurement of vacua, which has been the subject of the preceding chapters.

Many of the usual physical and chemical constants of materials are needed; as these are readily obtained from standard reference books, they are not generally included here. Accounts given are further restricted to those materials frequently of concern in the vacuum laboratory rather than in industrial applications.

Because the information required is often miscellaneous and not always readily given in any logical sequence, no apology is offered for adopting an alphabetical sequence; at least it has the merit of facilitating reference.

Aluminium. A ductile metal of low density with good electrical conductivity (resistivity: $2{\cdot}45 \times 10^{-6}$ ohm cm at 0°C) which is difficult to degas thoroughly because of its low melting point of 658°C.

For constructional purposes, one of the aluminium alloys (where a wide range with different properties is available) is chosen. Drawn and cast aluminium are used for making vacuum chambers, base plates, cover plates, and flanges, where lightness is an advantage but where the ultimate pressure exceeds 10^{-6} torr. The availability of tube and sheet which is readily worked and shaped, light, and of good thermal conductivity frequently leads to the choice of aluminium for making the nozzle systems of oil vapour pumps, but it is attacked vigorously by mercury.

Aluminium is easily evaporated by heating *in vacuo* and is widely employed for vacuum coating (p. 262) glass, plastics, and metals for mirror making. Aluminized, optical, front-surface mirrors have an overall white-light reflectivity of 88% for normal incidence, and also high reflectivity in the ultra-violet region (80% at 2,500 Å).

Permeation of gas through aluminium is small even when it is in the form of thin foil; with the additional advantage of its low atomic number, this leads to the use of aluminium windows to vacuum vessel for passage of X-rays and high energy electrons.

A thin film of oxide (Al_2O_3) forms on a fresh surface of aluminium as soon as it is exposed to the atmosphere and increases in thickness for several days. The thickness of this oxide film can be increased by anodizing (p. 212). It gives protection of an aluminized mirror, and also makes soldering difficult. Sputtering occurs at a very low rate because of this oxide film – an advantage in using aluminium electrodes in a gaseous discharge.

Aluminium wire has been used considerably for making seals between machined (surface finish: 20 microinch r.m.s. or better) stainless steel flanges in ultra-high vacuum systems. Super-purity (99·99%) 0·036 inch diameter aluminium wire is used with the ends butt-welded in a small gas flame, using aluminium welding flux to form an endless ring gasket. The gasket is first annealed in an oven at 500°C, cleaned, and then compressed to less than 0·013 inch between the flanges concerned by bolting them together using, for example, eight $\frac{5}{16}$ inch bolts on a P.C.D. of $4\frac{1}{2}$ inch in $\frac{1}{2}$ inch thick flanges. On baking under vacuum to 400°C and above, the aluminium partly flows plastically and adheres to the flanges (*see* Holden, Holland and Laurenson [96]). The use of Schnorr washers (see p. 175) is recommended on the bolts used.

Aluminium 'cooking' foil is cheap and useful as a heat reflector, temporary electrical shield, and even for making vacuum-tight seals between mating steel edges.

Anodizing. The resistance of a metallic surface to abrasion can be increased greatly by the technique of anodizing. The method is usually applied to aluminium, which can be in the bulk form with a good surface finish, or as a vacuum coated film on a substrate. A suitable electrolyte is 3% tartaric acid in distilled water, with ammonium hydroxide added to give a pH of 5·5. The aluminium is made the anode in the electrolyte, with an adjacent parallel sheet of 99·99% aluminium as the cathode. The electrical supply is at about 200 volt d.c. with a current density of about 2 amp per sq foot. The oxide film grows to a maximum thickness in about 2 min, and increases linearly with voltage above 40 volt. At the maximum, insulation provided by the oxide forbids further thickness increase. The mean rate of increase of oxide film thickness over the voltage range, 0 to 200 volt, during anodizing is about 13 Å per volt across the electrodes.

Anodizing can also be used to form a tough oxide film on tantalum, using an ammonium borate electrolyte (*see* Hass [111] and Holland [112]).

Asbestos. The natural mineral fibres are hornblende (melting point: 1,150°C) and serpentine (melting point: 1,550°C). Such fibres are combined with binders to give commercially available asbestos in the form of sheet, string, paper, and wool. Density is 2·5 gram per cu cm approximately. Widely used for thermal lagging (thermal conductivity: 3 to 6×10^{-4} cal per cm per sec per degC).

For bake-out ovens, heat insulating platforms, etc., the commercial asbestos-cement (e.g. Transite) or asbestos-slate mixtures are harder and more robust. Marinite Ltd. manufacture an easily machined material available in sheets called 'Marinite', which is composed of asbestos fibre, diatomaceous earth, and an inorganic binder able to withstand continuously temperature up to 530°C. It is useful for oven construction. 'Sindanyo' is an asbestos-cement board material manufactured by Turners Asbestos Cement Co., Ltd., which will withstand temperatures up to 350°C, and considerably higher values if some loss of mechanical strength is allowed for. It is a harder, more robust material than Marinite, but not so easily machined. It is a useful material for providing an insulating bench-top. Typical applications of Marinite and Sindanyo are shown in Fig. 99.

Bake-out Ovens. Two electrically-heated bake-out ovens for laboratory vacuum systems are shown in Fig. 99: (*a*) is a pattern which is raised or lowered over the vacuum chamber by a counterbalance pulley arrangement; (*b*) is a unit construction type erected around the vacuum chamber and readily sectionalized. The heater in (*a*) is in the form of a coil of 'Brightray' wire wound inside the oven, the wire being spaced uniformly over the whole inside area of the oven so as to maintain as uniform a temperature as possible. Alternatively, as shown in Fig. 99(*b*), commercial radiant heater elements are installed within the oven. A current of 20 to 25 amp at 240 volt gives some 5 kW of electric power, suitable for raising an oven of type (*a*) and of dimensions $2 \times 1 \times 1$ foot to about 650°C; where the normal highest bake-out temperature for stainless steel or borosilicate glass is 450°C. The more efficient oven of type (*b*) containing aluminium reflector walls will attain a temperature of 400 to 450°C with a power supply of 1 kW per $1\frac{1}{2}$ cu foot.

The electrical supply to the oven is normally a.c., controlled by a

Variac transformer to be able to regulate the temperature obtained. Bennett [113] discusses the heating power requirements more fully.

Gas heating is used in some industrial vacuum practice. The ring-shaped ovens, on vacuum systems for mass-producing radio valves

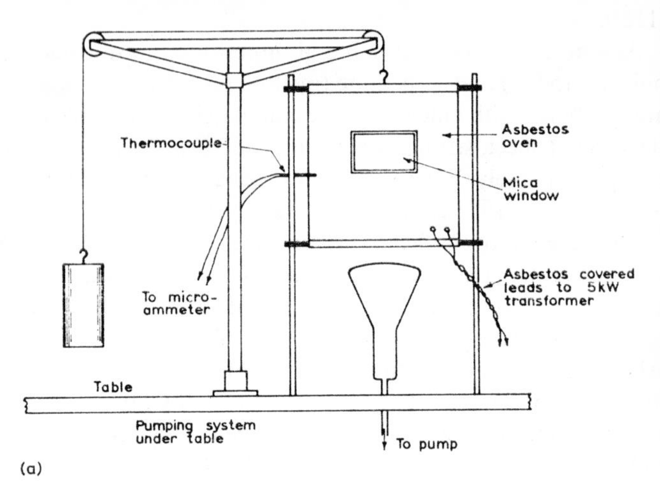

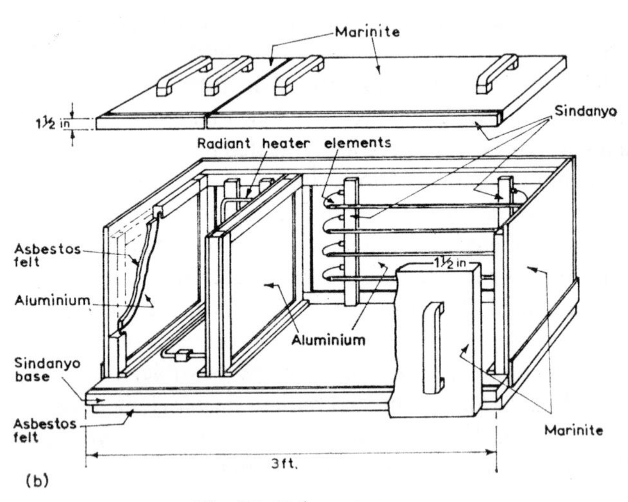

Fig. 99. Bake-out ovens.

on a turntable arrangement, are sometimes heated by numerous gas-jets arranged around the base of the inside wall of the oven, and provide a readily controllable means of obtaining uniformly dissipated heat at temperatures up to 1,000°C.

Brass. A copper-zinc alloy which is easily machined and extensively used in constructing scientific apparatus. But it is best avoided, because of gas permeability, in constructing vacuum chambers, connecting tubes, valves etc., especially if cast, though drawn tubing is tolerable at pressures above 10^{-5} torr. When heated *in vacuo*, brass readily sublimes zinc, the vapour pressure of which is 10^{-2} torr at only 343°C, so brass terminals to heater elements *in vacuo* are deprecated, copper being much superior.

Tombac is a corrugated flexible tubing available in various diameters and contains 28% zinc. It withstands heating to 450°C in air and is sufficiently non-porous to be useful for flexible connections at pressures down to about 10^{-5} torr. It is often used as tubulation to join the backing pump inlet to the vapour pump discharge outlet, but stainless steel corrugated flexible tubing is much superior.

Bronze. A copper-tin alloy. Castings are less porous than brass and have been used in fabricating rather complex vacuum chambers and sections of chambers, e.g. for ultra-violet vacuum spectrographs. It is, however, prone to air leaks, which can be avoided in some circumstances by heavily painting the surface, or coating it with a lacquer or W-wax.

Carbon. There are many forms of carbon, depending on the production process. Blocks of carbon are made from carbon powder by heating it to temperatures up to 3,000°C under pressure, and the mechanical properties obtained depend on the size and form of the powder particles used and the heating and sintering process.

Crystalline forms of carbon are *graphite* and, in perfect geometrical form, *diamond*. The density of graphite is 2·22 gram per cu cm and of diamond 3·51 gram per cu cm. The melting point of graphite is about 3,700°C. Resistivities at 20°C are 4 to 7×10^{-3} ohm cm for gas carbon, 3×10^{-3} ohm cm for graphite, and 10^{12} ohm cm for diamond. Another form of carbon is *charcoal*, prepared by heating alder wood, hornbeam, or coconut shell at 500 to 700°C in an iron container, so that it is destructively distilled until vapour evolution is insignificant.

It is difficult to degas carbon by heating *in vacuo*. The minimum temperature for removal of most of the sorbed gas is 1,800°C. Gases evolved are hydrogen, carbon monoxide, and a small amount of nitrogen. It absorbs oxygen on subsequent exposure to air.

Vapour pressure is 10^{-5} torr at 1,980°C, rising to 1 torr at 2,870°C.

Some applications in vacuum technology are:

(*i*) in the form of graphite (free of CO_2 and with a maximum of

0·5% inorganic impurities) electrodes for vacuum tubes, chiefly as anode of small transmitting valve where large heat dissipation is necessary and good heat conductivity and radiation are an advantage;

(*ii*) as a conducting coating on the inside of glass wall of vacuum tube (e.g. cathode ray tube), applied initially in the form of 'Aquadag';

(*iii*) as a gas sorbent.

Aquadag is a colloidal suspension of graphite in water or oil. This is applied to the glass (usually by means of a suitable brush) to form a black conducting film. It is then baked on to the glass at 450°C. For coating metal, it is best sprayed on from a gun and vacuum-baked at 900°C; greatest adhesion is obtained by coating onto a metal which has the same coefficient of thermal expansion. The addition of 3 to 4% by volume of sodium silicate to Aquadag improves its adhesion to glass. Typical applications are as electrical screening on the inside or outside walls of a tube, for making contact to metal coatings, to reduce secondary and photoelectric emission from electrodes, and as a coating on cooling fins (e.g. attached to thermionic valve grid supports). It absorbs caesium strongly. Also used to improve the electrical conductivity of paper, asbestos, and rubber by impregnation technique.

Carbon shaped into the form of a suitable jig or support is useful for glass-blowing and air brazing. For example, uniform heating in the making of glass-to-metal seals can be achieved by erecting the parts within a suitable carbon jig, and induction heating the carbon in air by a surrounding coil carrying radio-frequency current.

As described in section 1.15, activated charcoal in granular form is a useful gas sorbent, especially at −183°C (liquid oxygen) or −196°C (liquid nitrogen), which has been applied to the construction of sorption pumps but is at present giving way to the use of molecular sieve materials. These sorption characteristics are fully described by Dushman [114]. Table 5.1 gives typical figures.

TABLE 5.1
Sorption of gases by charcoal

Gas	*Volume in cu cm at s.t.p. of sorbed gas per gram of charcoal at*	
	0°C	*−183°C*
Helium	—	1·84
Hydrogen	1·55	100
Nitrogen	13	185
Oxygen	16	250

Cements, Adhesives, and Waxes. Included are details of several materials for sealing together flanges, cones, and other members, to make a semi-permanent or permanent joint or for covering a small hole or crack in a vacuum chamber. The use of one or other material depends primarily on the temperature it will have to withstand and the degree of permanency required.

Apiezon Q compound is a plasticine-like substance consisting of graphite mixed with the low vapour pressure residues of paraffin oil distillation products. It is a semi-solid material, which can be readily pressed with the fingers round a joint in glass or metal to render it free of leak. The vapour pressure is 10^{-4} torr at 20°C; it becomes too soft for use at temperatures above 30°C. Normally only used for making a temporary seal until such time as a permanent joint is effected, or in crude leak-detection practice.

An alternative is *silicone-putty*.

Apiezon W-wax is a hydrocarbon-based, hard, black substance supplied in sticks, with a vapour pressure below 10^{-8} torr at 20°C and 10^{-3} torr at 180°C. Softening point is 60 to 70°C. It is useful for making semi-demountable joints in glass or metal where fusing, soldering, or welding is not possible. It is applied to the precleaned surfaces to be joined at about 100°C. The surface must be heated as well as the wax if a satisfactory leak-free joint is to be made not liable to develop cracks. Over-heating of the wax will cause bubbling and charring. In making the joint, the surfaces should be moved over one another, if possible, to expel air. W-wax is soluble in xylene.

Apiezon W40 and W100 waxes have softening points at 30°C and 50°C respectively and vapour pressures similar to W-wax. They provide a less robust joint than W-wax, but are useful when the joint cannot be heated much.

Picein is hard black wax based on bitumen and has a vapour pressure of 10^{-8} torr at 20°C and about 5 torr at 50°C. Softening point is 50°C. Used similarly to W-wax but has higher vapour pressure at elevated temperatures. Chemically inert to usual organic liquids and inorganic acids.

Edwards waxes WE3 and WE6 are shellac-based, brown waxes which, unlike W-wax and Picein, are soluble in alcohol but fairly insoluble in aromatic hydrocarbons like benzene and toluene. Softening point of WE3 is 80°C and of WE6 it is 90°C. Used like W-waxes.

Khotinsky cement is a mixture of shellac and pitch obtained from Caroline tar which softens at 50°C. Comparatively insoluble in usual organic liquids and common acids. Useful for joints at temperatures below 40°C, but has vapour pressure of 10^{-3} torr at 20°C

Glyptal is a viscous, condensation resin which resists the action o
water, acids, and alkalis. Applied like a paint or varnish at room tem
perature, it dries to a hard glossy film in 8 hour at 20°C. Used fo
painting onto regions of porosity in walls of vacuum system or vessel

Sealing wax is soluble in alcohol and is convenient in solution fo
painting onto a leaky region as an alternative to glyptal.

Polyester and epoxy resins are obtainable in liquid form of viscosity
low enough to permit application by a brush. Preferred to glyptal a
outgassing rate is lower (Bailey [115]).

Silicone varnish DC 997 can be applied by brush or other means t
seal a small leak. It will withstand subsequent baking to 300°C, un
like glyptal and the resins, so is very useful for sealing a region o
small leakage rate in the wall of a vacuum chamber or system which
has to be baked, as in obtaining ultra-high vacua.

Silver chloride applied at 450°C is a useful sealing cement which wil
withstand high temperatures. Soluble in sodium thiosulphate. Can b
used for sealing small flat windows onto glass vessels and for sealing
metal leads into discharge tubes. Slow cooling is essential to preven
cracking.

Araldite (Ciba (A.R.L.) Ltd.) is an epoxy resin adhesive of high
strength, low outgassing rate, and low shrinkage during set.

Thermo-setting Araldite 1 is available as sticks or powder, eithe
light-brown or silver. It is most useful for joints between well-fitting
surfaces, but may also be set in a small well between, for example, a
lead-in and a surrounding ring. The surfaces to be joined may be o
glass, metal, ceramic, quartz, or mica; thorough precleaning i
necessary.

To use Araldite 1 powder, it is sprinkled over the clean surfaces an
they are then bonded by the application of heat alone, without pres
sure. Suggested curing temperatures and times are 240°C for 10 min
220°C for 20 min, 190°C for 60 min, or 180°C for 120 min. If stick
form Araldite is used, the surfaces to be joined must first be heated t
120°C, the stick rubbed on, and a curing operation then employed a
for the powder.

Seals made with thermo-setting Araldite can withstand tempera
tures up to 150°C under vacuum, but prolonged use above thi
temperature causes gradual decomposition of the Araldite.

An ethoxylene resin Araldite 101 with hardener 151 sets at roon
temperature and has a low softening point, so that above 60°C th
joint can be dismantled.

Bailey [115] strongly recommends for vacuum work the Ciba epoxy resin AY 111 as a cold-curing two-part adhesive, which has a short pot life of only 30 min, but is capable of making strong vacuum joints without heating during application being necessary.

The chief material evolved from Araldite 1 in vacuum is water vapour, as shown by mass spectrometer study (Turnbull, Barton, and Rivière [116]).

Ceramics. Inorganic non-metallic compounds of a crystalline nature, such as pure oxides, mixed oxides, silicates, aluminates, titanates, borides, carbides, and others, are fired so as to produce ceramics. These are essentially crystalline aggregates, but where the crystal forms usually undergo considerable changes from the initial states during firing, and also an amorphous or 'glass' phase is present, the extent of which has a considerable influence on the ceramic produced. *Cermets* are formed by compounding metal powders with the ceramic batch to give an electrically-conducting material, with enhanced oxidation resistance compared with the refractory metal. Glass-ceramics such as *Pyroceram* are essentially special glasses which are converted into ceramics by adding nucleating crystal-forming agents to the glass batch before firing.

Only the ceramics frequently encountered in vacuum laboratory techniques are briefly described below. The field demands much wider study in its application to the design of special electron tubes, vacuum furnaces, and space missiles. Kohl [117] gives comprehensive accounts.

Aluminium oxide or alumina (Al_2O_3) is the most widely used ceramic in vacuum techniques. The pure crystalline form of aluminium oxide is sapphire. Alumina is available in five main grades with the percentages of Al_2O_3 denoted: alumina A (99% Al_2O_3); alumina B (97%); alumina C (96%); alumina D (94%); alumina E (85%).

Alumina has good mechanical strength, is impermeable to gases, and is unaffected at temperatures up to 1,700°C by air, water vapour, vacuum, or permanent gases. Sintered alumina (99·8% Al_2O_3) has a fusion temperature of 2,030°C, a thermal conductivity of 0·069 cal per cm per sec per degC, which decreases to 0·014 at 1,000°C, and good resistance to thermal stress.

Ceramic envelope vacuum tubes, such as high power triodes, pulse triodes, short-wave thermionic tubes, klystrons, magnetrons, rectifiers, and travelling-wave tubes, most commonly have this envelope made of high purity alumina ($>85\%$ Al_2O_3).

Alumina-metal high vacuum lead-ins and terminals are particularly useful for large currents. The alumina and the metal sealed together should have the same thermal expansion coefficient, so Nilo K or Kovar is often used as the metal component. However, it is not possible to obtain thermal matching over the range of temperatures from 0 to 550°C encountered in use and bake-out, so the seal is designed to ensure that the ceramic is under compression, as tension would lead too readily to fracture.

Alumina electrode spacers for electron tube assembly provide rigid support with excellent electrical insulation, ideally clean and readily degassed. Lathe working and drilling are difficult because of the hard brittle nature of alumina; grinding and polishing are accomplished, however, and mouldings in a variety of shapes are commercially available.

The insulation between the heater wire (often of molybdenum-tungsten alloy) and the surrounding cathode of an indirectly-heated oxide-coated cathode is frequently in the form of alumina rod containing suitable bore holes in which the heater wire is threaded; alternatively, alumina powder in colloidal suspension in amyl acetate is sprayed onto the heater wire from a gun and subsequently dried in air.

Beryllia (BeO) is similar to alumina in its resistance to gases and vacuum on heating to 1,700°C, but it volatizes quickly in the presence of water vapour because beryllium hydroxide is formed. Its thermal conductivity is 0·5 cal per cm per sec per degC at 100°C, decreasing to 0·046 at 1,000°C for 99·8% BeO, considerably greater than for alumina. This higher thermal conductivity leads to the use of beryllia instead of alumina in electron tube structures such as supports for the helices of travelling-wave tubes and microwave windows for high power klystrons, where design for high heat dissipation is important. Grinding and firing of beryllium oxide produces a toxic dust or vapour which is dangerous.

Magnesia (MgO) has poor thermal stress resistance, mechanical strength and reacts in contact with metals at elevated temperatures. It has been used as a sprayed-on insulating coating for heater wires in indirectly-heated cathodes but, in general, is not recommended for use in vacuum. It has a thermal conductivity of 0·082 cal per cm per sec per degC at 100°C, decreasing to 0·016 at 1,000°C.

Zircon is a combination of zirconium oxide (ZrO_2) and silica (SiO_2) with good heat shock resistance, and has been used considerably for making vacuum furnace trays. Its thermal conductivity is 0·009 cal per cm per sec per degC at 100°C.

Steatite in the form of 'block talc' or 'soap-stone', a natural steatite, has the considerable advantage that it is easily machined, which is not possible with the other oxide ceramics described above. After machining, it must be fired to hardness in air before use, when shrinking will take place. Though an infrequent present-day choice as a ceramic, it is often useful in the laboratory because of the ease with which it can be made into required shapes on the lathe or milling machine. Steatite, as a compound of magnesium oxide (MgO) and silica (SiO_2), is also manufactured, particularly as an insulating material with low dielectric loss at high frequency. Steatite in the form of Alsimag 228 has a thermal conductivity of 0·006 cal per cm per sec per degC at 20°C. Enrichment with magnesium compounds in manufacture gives the ceramic *Forsterite* ($2MgO.SiO_2$), having far superior electrical properties at microwave frequencies and with a thermal expansion coefficient sufficiently linear over a wide temperature range to enable sealing to titanium and to nickel-chromium-iron alloys.

Boron nitride (BN) is an alternative to natural steatite in that it can be readily machined and with the advantage that subsequent firing is not needed.

Mullite (also Sillimanite) is a mixture of alumina and silica ($3Al_2O_3.2SiO_2$) which has inferior properties mechanically and electrically to the oxide ceramics described. An advantage in vacuum practice is that Mullite can be sealed to certain hard glasses.

Ceramic-to-metal sealing has been undertaken by a variety of techniques in making electron tube envelopes, electric lead-ins, and terminals for vacuum chambers. A comprehensive account is given by Kohl [117]. Two techniques reasonably easily undertaken in the average laboratory are, first, to apply a metallizing coating of platinum to the ceramic by painting on a platinum metallizing preparation (*see* p. 252) with a camel-hair brush, firing to 650 to 800°C and then soldering the metal to the platinum film, and, second, to use the Bondley technique.

The Bondley [118] method is a zirconium hydride (ZrH_4) process in which the scrupulously cleaned ceramic is coated with a thin layer of paste made from powdered zirconium hydride in water or a nitrocellulose binder. A ring or other suitable shape of silver-copper eutectic solder or pure silver is placed in contact with this layer and sandwiched between it and the fitting metal part. The assembly is heated to 1,050°C in a vacuum stove at a pressure below 10^{-4} torr. The zirconium hydride disintegrates before the solder melts, the hydrogen evolved is pumped away and the zirconium left bonds strongly to the ceramic and alloys with the silver soldered to the metal.

A useful ceramic-metal seal which can be made by the zirconium hydride process is alumina (98% Al_2O_3) to Nilo K, copper, or nickel. For the many other possibilities, see Kohl. [117].

Copper. Excellent thermal and electrical conductivity (thermal conductivity: 0·144 cal per cm per sec per degC at 18°C; electrical resistivity: $1{\cdot}78 \times 10^{-6}$ ohm cm at 18°C) and a melting point of 1,083°C lead to use of copper for construction of nozzle systems of oil vapour pumps, cooling coils, and cooling fins, specialized vacuum tube electrodes, and, as the gas permeability is small, for constructing vacuum chambers. Amalgamates with mercury.

Housekeeper [119] seals between copper and glass are made in tubes up to 4 inch diameter, the end of the copper tube being bevelled to a 'feather' edge to take advantage of the ductibility of copper in avoiding undue stress on the glass when heating.

A nickel-iron-alloy wire thinly coated with copper can be readily sealed into most glasses to give a vacuum-tight lead-in wire. The wire is usually treated with a thin layer of borax flux to facilitate sealing to the glass. Known as 'red-platinum' wires because of their ability like platinum to seal to glass, these wires are used as pinch-wires in lamp bulbs and also for radio receiver valves of a now largely obsolete type.

Safe currents for copper are 270 amp per sq cm for 12 s.w.g. wire of diameter 2·64 mm, 430 amp per sq cm for 22 s.w.g. wire of diameter 0·711 mm, and 500 amp per sq cm for smaller diameters.

Copper vacuum chambers are largely impermeable to gas and are convenient for large-scale work, though mild steel is more usual as it is less expensive, and stainless steel is preferred for ultra-high vacua. Tinning or nickel-plating the copper surface, where plating is especially good practice for the surface exposed to the vacuum, is recommended because pure copper corrodes and oxidizes readily in the atmosphere.

To remove the heavy oxide from copper, it is dipped in warm inhibited 75% hydrochloric acid; the inhibitor being Rodine No. 50 in 0·25% by volume. It is subsequently rinsed in water and dried.

Oxygen-free, high conductivity (O.F.H.C.) copper is preferred for use *in vacuo*, especially for electrodes, sealing wires, tubes, and gaskets, because ordinary electrolytic copper contains oxide inclusions which lead to the formation of water vapour, causing small cracks, porosity and brittleness when copper is hydrogen-stoved.

Copper is readily soft-soldered and brazing is best with silver-copper eutectic alloy.

O.F.H.C. copper gasket seals between mating flanges, usually of stainless steel, are generally of diamond cross-section (Turnbull, Barton and Rivière [116]) or in the form of flat washers between 60° knife-edges on the flanges themselves. Compression to about 85% of the original apex-to-apex thickness by means of stainless steel flange bolts is recommended for the diamond-section copper gasket. The linear expansion coefficient of copper ($16{\cdot}7 \times 10^{-6}$ per degC at 20°C) is conveniently near that of stainless steel, so the copper gasket seal can be heated to 450°C and cooled without the flange bolts becoming slack. For a permanent copper gasket seal, it is recommended by Turnbull *et al.* to insert the diamond-section gasket ring in a square-section annular channel of depth equal to 85% of the apex-to-apex gasket thickness in one flange, and then bolt down the other flat flange to give the recommended compression when it mates with the outer flat surface of the channeled flange. Eight $\frac{5}{16}$ B.S.F. clamping bolts of stainless steel on a pitch-circle diameter of 3 inch are recommended in the flanges with a diamond-section copper gasket apex-to-apex thickness of $\frac{3}{8}$ inch.

Degassing. To attain a pressure in a chamber below 10^{-7} torr, it is, in general, essential to bake the chamber to degas its walls, even if the chamber is continuously pumped. For vessels which are subsequently to be sealed-off from the pump and also where, whilst pumping, pressures of 10^{-9} torr and below are to be maintained, this bake-out to degas the chamber has also to be accompanied by rigorous degassing of materials and assemblies within the chamber. Such degassing of interior parts will, of course, occur during bake-out. However, degassing is more rapid and more effective the higher the temperature, so metal parts are best degassed before assembly.

Glass vessels undergoing pumping are usually baked at temperatures not exceeding 350°C for soda-lime and lead glasses, and 450°C for borosilicate glasses such as Pyrex and Hysil. Stainless steel chambers, frequently employed in ultra-high vacuum technology, can be baked to 450°C.

Thermal tapes and electric heating mantles are convenient for baking selected parts of a vacuum system.

Metal and graphite electrodes and other parts can be degassed by heating *in vacuo* to temperatures not exceeding those listed in Table 5.2.

A vacuum stove suitable for degassing metal and carbon parts

TABLE 5.2

Degassing temperatures for metals and graphite and gases evolved given in order of decreasing amounts

Material	*Maximum temperature* °C	*Gases evolved*
Copper	500	CO_2, CO, H_2O, N_2, H_2
Graphite	1,800	N_2, CO at 2150°C
Iron	1,000	CO, H_2O, N_2
Molybdenum	950*	CO, H_2, N_2
Nickel	950	CO, H_2O, CO_2, H_2
Platinum	1,000	
Steels	1,000	CO, H_2, N_2, CO_2
Tantalum	1,400	CO, H_2, N_2, H_2O
Titanium	1,100	H_2, H_2O, CH_4
Tungsten	1,800	CO, CO_2, H_2

* Molybdenum can be heated to 2,150°C to degas it thoroughly but it becomes brittle if heated to above 950°C *in vacuo*.

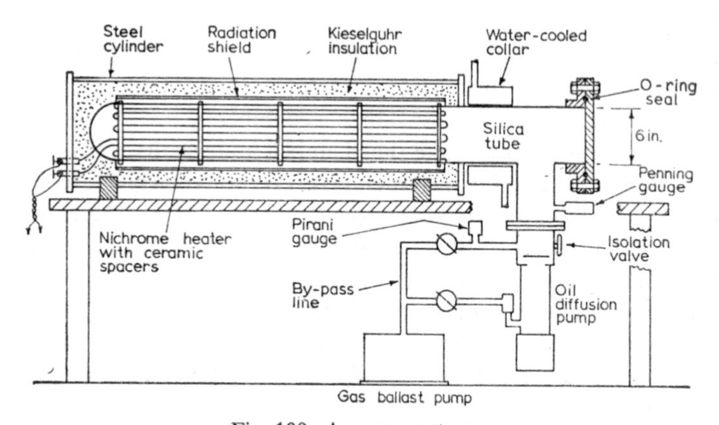

Fig. 100. A vacuum stove.

(Fig. 100) is constructed with a silica or ceramic cylindrical chamber and is pumped to a pressure of 10^{-4} torr or below by an oil diffusion pump. This pump needs to be adequately baffled to minimize oil back-streaming into the vacuum stove; in cases where stringent avoidance of such back-streaming is essential, a liquid-nitrogen-chilled cold trap between the oil diffusion pump inlet and the cylindrical chamber is desirable.

Some metals, in particular, molybdenum, platinum, nickel, and iron, may be heated in a hydrogen atmosphere to clean them rather than in a vacuum stove. If required, hydrogen-stoving can be followed by vacuum-stoving, but hydrogen sorbed is re-evolved in the bake-out at 450°C on the final vacuum system.

Instead of using electrical resistance heating, an induction-heated vacuum stove may be used in which, around the pumped glass or silica tube, is placed an induction coil (eddy-current heater coil) carrying radio-frequency current. The metal parts within the evacuated tube inside the surrounding coil are then heated by eddy-currents induced in them (*see* p. 244).

Degreasing agents. To ensure that the surface of a metal is free of grease, the procedure adopted is rinse in acetone, rinse in three changes of trichlorethylene, then in two changes of methyl alcohol, and dry in a blast of warm air or in an oven at 70 to 110°C. Alternatively, an excellent method is to degrease metal in an isopropyl alcohol vapour degreaser, as used for cleaning glass optical components. Holland (private communication) reports that a high coefficient of friction of aluminium wire on stainless steel is a good guide to surface cleanliness before baking.

Drying Agents. Of the drying agents listed in Table 5.3, the ones

TABLE 5.3

Drying agents

Drying agent	*Mass in mg of water remaining per litre of gas at 25°C after drying*
Charcoal trap at −196°C	$1{\cdot}6 \times 10^{-22}$
Phosphorus pentoxide	5×10^{-5}
Magnesium chlorate (anhydrous)	$1{\cdot}6$ to $2{\cdot}4 \times 10^{-3}$
Calcium oxide	3 to 4×10^{-3}
Silica gel	2×10^{-3} to 10^{-2}
Caustic potash (sticks)	10^{-2} to $1{\cdot}7 \times 10^{-2}$
Zinc chloride (sticks)	0·94 to 1·02
Calcium chloride (anhydrous)	1·23 to 1·27
Calcium chloride (granular)	1·4 to 1·6
Copper sulphate (anhydrous)	2·7 to 2·9

See Bower [120] for further information.

most used in vacuum technique are phosphorus pentoxide, to assist in reducing the partial pressure of water vapour inside the evacuated system (section 3.5), and silica-gel, to dry the air admitted to a vacuum system when bringing it up to atmospheric pressure. Water vapour in admitted air will become sorbed on the inside walls of a vacuum chamber and so prolong the subsequent pump-down time to the ultimate pressure required. In ultra-high vacuum technology, the admission of dry air is especially advantageous; indeed, when using cold-cathode getter-ion pumps, much time is saved in ultra-high vacuum pumping schedules if dry nitrogen is admitted. Silica-gel, which can be replenished by baking in air, unlike phosphorous pentoxide, is used, in a container in the air or nitrogen admittance line.

Elastomers. *See* Rubbers, p. 252.

Electrolytic Cleaning and Polishing. Several metals can be cleaned electrolytically by arranging the metal part to be the *anode* in an electrolytic cell and deplating it by transferring metal from its surface to a suitable cathode. During this process, some etching takes place, so the method is not recommended for thin tungsten and molybdenum filaments and ribbons intended for subsequent use *in vacuo* as thermionic emitters. The surface also becomes electropolished, especially with metals such as copper, brass, and stainless steel, because prominences and points on the surface are preferentially removed in the deplating.

Any metal part to be cleaned or polished electrolytically is first degreased.

To clean tungsten or molybdenum electrolytically, a suitable electrolyte is 20% potassium hydroxide (KOH) solution, with the tungsten or molybdenum as the anode and a carbon cathode. Filaments of these metals are cleaned in about 30 sec using a 7·5 volt, 200 watt d.c. supply. As the cathode is carbon, a.c., conveniently supplied by a transformer, can be used instead of d.c.

To clean nickel, the electrolyte used consists of 308 ml of water, 172 ml of sulphuric acid, and 546 ml of phosphoric acid, with the nickel as anode and a carbon cathode. The current density is adjusted until the nickel anode gases, but caution is essential because the solution dissolves nickel.

Electropolishing is now an accepted technique for the interior surface treatment of copper and stainless steel chambers and components to be used in ultra-high vacuum systems. It gives a bright, smooth, clean finish much preferred to that obtained by mechanical buffing or polishing which is likely to introduce undesirable inclusions of fine polishing powder within surface crevices. Copper components are best nickel- or chromium-plated if to be used in ultra-high vacuum.

After degreasing and often preliminary mechanical polishing of the metal surface, suitable electropolishing procedures are the following.

For copper, brass, nickel, and steel, an electrolyte consisting of 25 to 60% concentrations of orthophosphoric acid in water is used at room temperature.

TABLE 5.4

Physical data for some gases

Gas	*Molecular weight*	*Density at s.t.p. gram per litre*	*Boiling point °C*	*Critical temperature °C*	*Critical pressure atm*	*Viscosity at 760 torr and 20°C micropoise*	*Mean molecular diameter 10^{-8} cm*	*Mean free path at 10^{-3} torr and 20°C cm*	*Mean molecular velocity at 0°C 10^4 cm per sec*	*First ionization potential in volts*	
										For atom	*For molecule*
Air	—	1·2928	—	−140·7	37·2	181	—	5·1	4·47	—	—
Argon	39·944	1·784	−185·7	−122	48	221·7	2·90	5·07	3·81	15·68	—
Bromine	159·83	7·139	58·78	302	131	154	—	—	—	11·8	12·8
Carbon monoxide	28·01	1·250	−190	−139	35	175	3·15	—	—	—	14·1
Carbon dioxide	44·01	1·977	−78·5	31·1	73	148	3·28	3·34	3·62	—	14·4
Chlorine	70·91	3·214	−34·6	144	76·1	132·7	—	3·47	—	12·95	13·2
Helium	4·003	0·1785	−268·9	−267·9	2·26	194·1	2·30	14·6	12·01	24·46	—
Hydrogen	2·016	0·0899	−252·8	−239·9	12·8	87·5	2·36	9·3	16·93	13·527	15·6
Krypton	83·70	3·708	−152·9	−63	54	250	3·20	4·05	—	13·93	—
Mercury vapour	200·61	—	356·58	—	—	—	3·0	6·3	1·7	10·39	—
Neon	20·18	0·9002	−245·9	−228·7	25·9	311·1	2·6	10·4	5·36	21·47	—
Nitrogen	28·02	1·2506	−195·8	−147·1	33·5	176	3·2	5·1	4·54	14·48	15·51
Oxygen	32·00	1·429	−182·96	−118·8	49·7	202	2·95	5·4	4·25	13·55	12·5
Water vapour	18·016	0·606	100	374	217·72	98	4·60	3·4	5·67	—	12·56
Xenon	131·30	5·851	−107·1	16·6	58·2	226	3·7	3·0	—	12·08	—

With the copper or brass part as the anode and a suitable cathode (carbon or nickel), a current density of 3·5 amp per sq inch with a d.c. supply of 2 volt is generally used, electropolishing being complete in about 10 min. Several alternative electrolytes have been proposed and some are the subject of patents. An example is 59% concentrated phosphoric acid, 4% concentrated sulphuric acid, 1·5% chromic acid, and 35·5% water.

For stainless steel, electrolytes which have been used are: (*a*) 60% phosphoric acid, 20% sulphuric acid, and 20% water, with a current density of 5 amp per sq inch at 60°C; (*b*) 15% phosphoric acid, 60% sulphuric acid, 10% chromic acid, and 15% water, with a current density of 4 amp per sq inch at 50°C.

Electropolishing of stainless steel is easily carried out if the parts are of simple geometry, but special electrode configurations are needed to electropolish the interior of a stainless steel tube, and the technique then demands considerable experience to be successful. The cleaning and polishing are normally complete in about 30 min.

After electrolytic cleaning or polishing, the metal part must be rinsed in hot water, then in cold distilled water, and finally dried in warm air.

TABLE 5.5

Vapour pressures of some gases at various temperatures*

Gas	*Temperature in °K for vapour pressure given in torr*			
	10^{-12}	10^{-8}	10^{-4}	1
Ammonia	74·1	90·6	116·5	163
Argon	21·3	26·8	35·9	54·4
Carbon monoxide	21·5	26·7	—	—
Carbon dioxide	62·2	76·1	98·1	137·5
Helium	—	—	—	1·0
Hydrogen	2·88	3·71	5·38	9·55
Methane	25·3	32·0	43·5	67·3
Neon	5·79	7·34	10·05	15·8
Nitrogen	19·0	23·7	31·4	47·0
Oxygen	22·8	28·2	36·7	54·1
Water vapour	118·5	144·5	185	256
Xenon	40·5	50·8	68·1	103·5

* For further information see Honig and Hock. [121]

The partial pressures in torr of gases in the atmosphere at s.t.p. at sea-level are approximately: nitrogen, 590; oxygen, 160; argon, 7·05; water vapour, 5 to 30; carbon dioxide, 1·5 to 3×10^{-1}; neon, $1{\cdot}36\times10^{-2}$; helium, 4×10^{-3}; krypton, $8{\cdot}4\times10^{-4}$; hydrogen, $3{\cdot}8\times10^{-4}$; xenon, $6{\cdot}1\times10^{-5}$; ozone, $1{\cdot}52\times10^{-5}$; radon, $5{\cdot}3\times10^{-17}$. (*See Smithsonian Physical Tables*, 1954.)

Gases. Included are vapours, where a vapour is defined as a gas at a temperature below its critical temperature. A vapour can thus be condensed to the liquid or solid state by the increase of pressure above.

One mole of an ideal gas at s.t.p. (0°C and 760 torr) occupies a volume of 22·415 litre. The number of molecules per litre at a pressure of 1 torr and a temperature of 0°C is therefore $3{\cdot}54\times10^{19}$. Hence, the number of molecules per litre at a pressure of 10^{-n} torr is $3{\cdot}54\times10^{19-n}$.

Gases permeate through solid materials. The gas is first adsorbed at and then dissolved in the external surface of the wall of an evacuated enclosure. It then diffuses through the material of the wall in accordance with Fick's law to reach the interior wall, from which it is desorbed into the evacuated space (Norton [122]). The permeation process increases exponentially with temperature. Fick's law states that the rate of diffusion is inversely proportional to the thickness of the material at constant pressure difference and temperature.

For non-metals and with metals for monatomic gases, the rate of permeation of gas is given by

$$Q = \alpha P\frac{p_2-p_1}{d} \tag{5.1}$$

where Q is the permeation throughput in torr litre per sec, α is the cross-sectional area of a uniform slab of the material of thickness d, p_2 is the gas pressure on one side of this slab and p_1 that on the other, where $p_2>p_1$.

In the case of diatomic gases permeating through metals, dissociation of the molecules into atoms usually takes place, and the equation that applies is not (5.1) but

$$Q = \alpha P\left[\frac{\sqrt{p_2}-\sqrt{p_1}}{d}\right] \tag{5.2}$$

P, the permeation constant, depends on the gas and the metal, and is usually quoted in the unit torr litre per sec for a cross-sectional area of 1 sq cm, a wall thickness of 1 mm and a gas pressure difference of 10 torr or 760 torr. At room temperature, for materials used in the construction of vacuum systems, P is generally very small, so the time unit hour is often used instead of sec. However, P increases with the absolute temperature T approximately in accordance with $\exp(-1/T)$. Fig. 101 shows graphically the variation with temperature of P in torr-litre per hour mm per sq cm, for a pressure

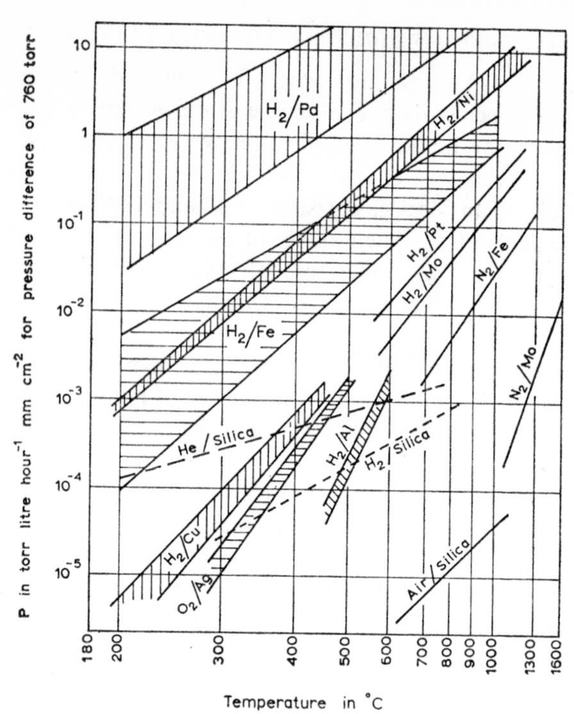

Fig. 101. The variation of the permeation constant for various gases and materials with temperature.

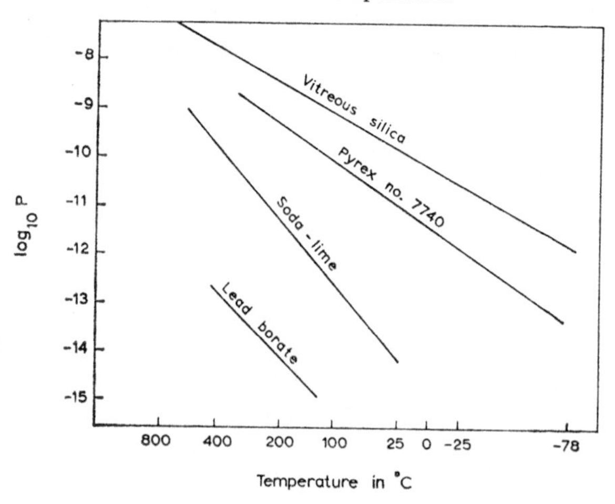

Fig. 102. Rate of permeation of helium through various glasses at various temperatures.

difference of 760 torr, for a number of combinations of gases and metals, and for helium, hydrogen, and air through silica (*see* Waldschmidt [123]; Espe [124]; Eschbach [125]).

Norton [1221] gives the helium permeation rate through various

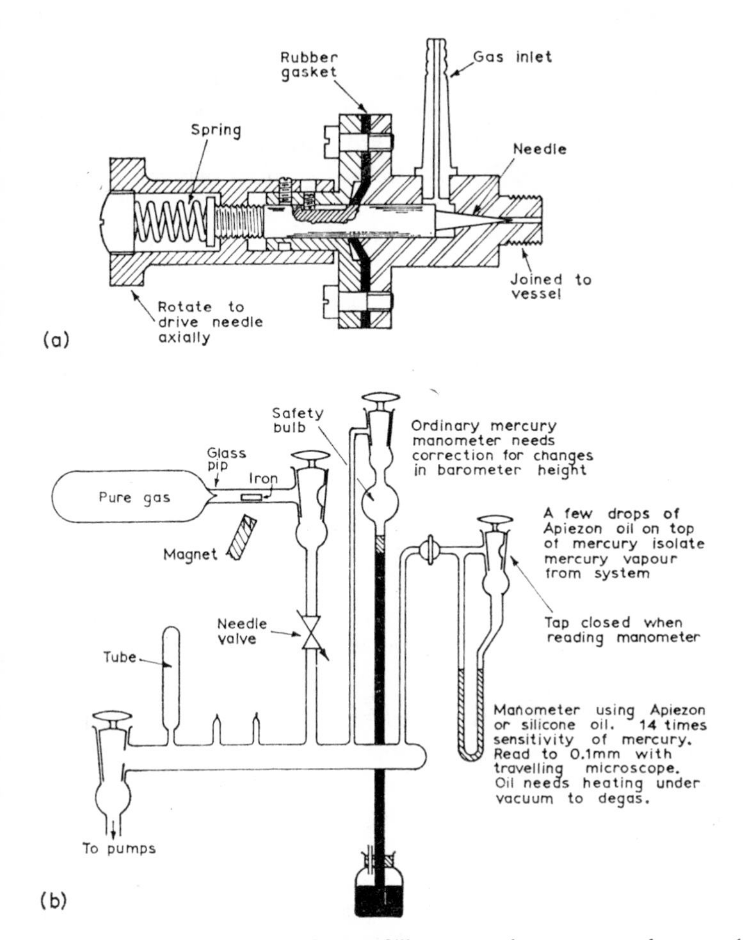

Fig. 103. (*a*) A needle valve. (*b*) Gas-filling a vessel to pressures between 1 and 500 torr.

glasses at various temperatures (Fig. 102). Note the comparatively high rate through borosilicate glasses such as Pyrex. This is a limiting factor in the region of 10^{-11} to 10^{-12} torr to the attainment of ultra-high vacua in a borosilicate-glass envelope on an ultra-high vacuum system where the available pumping speed is below 1 litre per sec, because of the permeation through the wall of helium in the atmosphere, where it has normally a partial pressure of 4×10^{-3} torr.

Gas-metering Methods. It is often needed to meter a gas into a vacuum at a controlled, small rate, so as to fill the previously evacuated chamber to a measured pressure with a known gas; examples being in the manufacture of gas-filled tubes such as thyratrons, cold-cathode trigger tubes, Geiger-Müller counter tubes, and in the measurement of pumping speed. Fig. 103(*a*) shows a needle valve suitable for this purpose; Fig. 103(*b*) shows a system incorporating such a valve for gas-filling a vessel to pressures between 1 and 500 torr. The ultra-high vacuum valve of the bakeable type (Fig. 80) is a good choice for admitting gas to an ultra-high vacuum system; the

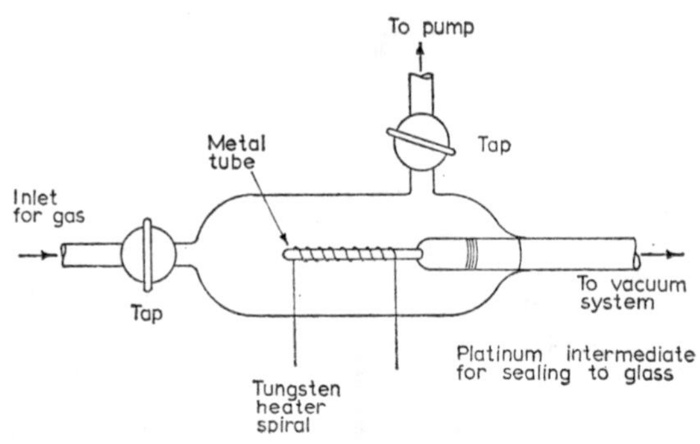

Fig. 104. Introducing gas into a system (*after* Jossem).

use of two such valves separated by a stainless steel or glass tube enables the pressure in a chamber, due to admission from a gas reservoir, to be raised from, say, 10^{-10} to 10^{-9} torr. An arrangement based on the permeation of hydrogen through heated palladium at about 900°C (*see* Fig. 101) due to Jossem [126] (Fig. 104) can be modified for admitting other gases, by making use of the diffusion of oxygen through silver, nickel, or copper; of nitrogen through molybdenum, iron, or chromium; of carbon dioxide through iron or nickel; and of a clear fused silica tube for the admission of helium and other rare gases. Joining the metal used to the glass bulb can be accomplished by an intermediate platinum tube, which seals directly to glass and is silver-soldered to the metal tube concerned, whereas, if silica is used, a graded glass seal is necessary.

Getters and Gettering. A getter is often included within a chamber which has been initially pumped to a low pressure by conventional means. The purpose of the getter is to sorb gas and retain it, so providing a method of reducing the gas pressure still further. Moreover, a wide use of getters is within sealed-off evacuated electron, or other, tubes, to sorb and retain gases evolved by the walls and electrodes so preventing rise of pressure, and, indeed, to reduce the pressure below that attained during conventional pumping before seal-off.

Three classes of getter are used.

(*a*) **Flash getters,** which are chemically active, electropositive, volatile metals (barium, magnesium, calcium, aluminium), evaporated by heating in the residual gases within a chamber previously pumped by conventional means – usually a vapour and/or rotary pump. On heating the flash getter metal *in vacuo*, the volatilization causes **dispersal gettering**, in which gas is sorbed by chemical action with the vaporized metal; the metal and any chemical salts of it which are formed then become deposited in thin film form on adjacent cold surfaces (usually the chamber wall). This film may then sorb gas by **contact gettering** because some of the gas molecules incident upon it become attached to the film, and may diffuse into its interior; the molecules, depending on the nature of the gas, have a certain sticking coefficient to the metal film.

(*b*) **Bulk getters,** which are filaments or ribbons of the getter metal, which are heated so that they sorb gas.

(*c*) **Coating getters,** consisting of non-volatile metal powders, which are coated on the surface of an electrode or other part in a vacuum tube or on the inside wall of the metal envelope of a tube.

The most widely used flash getter consists of a *barium-aluminium alloy*. A few mg of this alloy is packed by a specialized manufacturing process within a channel in a straight or ring-shaped iron or stainless steel tube (Fig. 105*a*). Several sizes are available. These getters are used within electron tubes, in particular, thermionic valves and cathode ray tubes; Fig. 105(*b*) shows a ring-getter positioned within the top of a miniature radio valve.

The getter is 'fired' after the processed and activated electron tube has been sealed-off from the pumps, or, in some cases, as the last stage in processing just before the final seal-off. The getter is heated by an induction-coil placed around the tube envelope (Fig. 105*b*), the ring-shape of the getter mount being particularly suitable as large eddy-currents are induced in it due to its coupling with the induction-heater coil. Alternatively, a straight-wire getter mounted between terminals may be fired by the passage of current directly through it.

The barium-aluminium getters are stable at temperatures up to 600°C and can be exposed to air without significant deterioration, though this is undesirable over long periods. The firing temperature

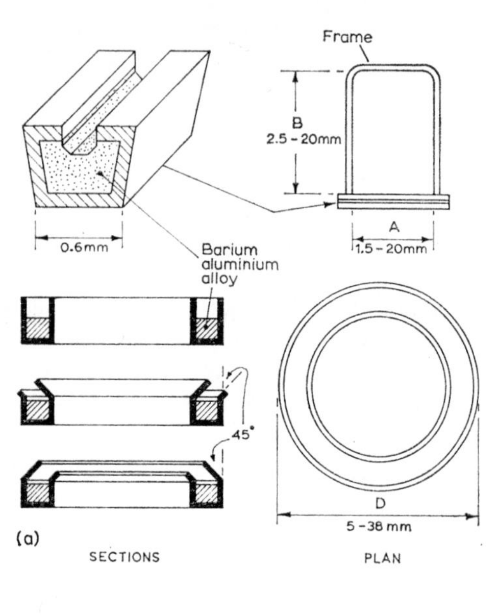

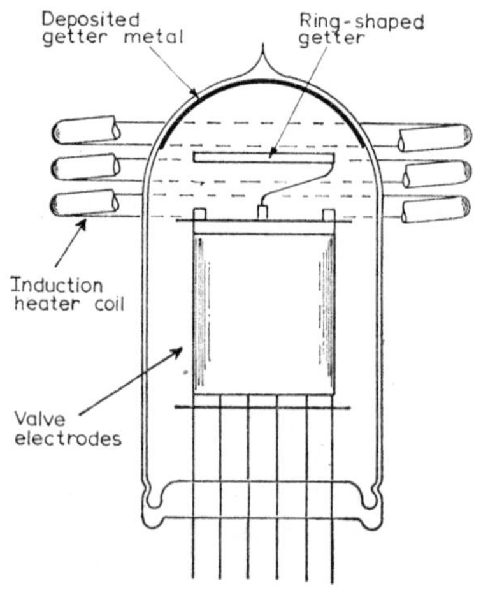

Fig. 105. Barium-aluminium alloy getters (Società Apparecchi Elettrici e Scientifici, Milan).

for effective dispersal *in vacuo* is 1,100°C. They getter effectively air oxygen, water vapour, hydrogen, carbon oxides, nitrogen (Table 5.6),

TABLE 5.6

Total sorption capacity of 1 mg of barium-alloy ring getter (della Porta [127])

Dry gas	*Quantity sorbed in torr-litre*
Air	$2{\cdot}6 \times 10^{-3}$
Carbon monoxide	$3{\cdot}7 \times 10^{-3}$
Carbon dioxide	$1{\cdot}8 \times 10^{-3}$
Hydrogen	13×10^{-3}
Nitrogen	$2{\cdot}25 \times 10^{-3}$
Oxygen	50×10^{-3}
Water vapour	61×10^{-3}

and other gases except the noble inert gases, which are only adequately contact gettered if they are ionized.

The firing of a barium-aluminium getter in an electron tube sealed off initially at a pressure of 10^{-3} torr can reduce the pressure to 10^{-9} torr.

Amongst several other metals used or which have been used in the past for dispersal gettering, only titanium and zirconium are frequently employed, usually in getter-ion or getter pumps (section 1.14) where the metal (usually titanium) is dispersed either by evaporation or by sputtering. These two metals are also useful as bulk getters. Titanium, which has a melting point of 1,660°C, is initially degassed by heating *in vacuo* to about 1,300°C to drive off dissolved hydrogen. A filament or sheet or layer in powder form of degassed titanium heated to above 400°C (operating temperature in the range 500 to 1,100°C) will sorb hydrogen rapidly, and at a slower rate will sorb the gases oxygen, nitrogen, carbon monoxide, and carbon dioxide.

Titanium hydride powder has also been used as an effective getter in some types of thermionic vacuum tube with metal envelopes: on heating on the pumps, the hydrogen is driven off and pumped away, and the titanium remaining is the active getter.

Zirconium (melting point: 1,860°C) is a rather less efficient alternative to titanium; it absorbs hydrogen to an extent of about 1,500 times its own volume at s.t.p. when heated to 300 to 500°C but re-evolves it at 800°C or more. Though oxygen and water vapour are effectively sorbed also at 300 to 500°C, a temperature of 800°C is necessary for nitrogen, carbon dioxide, and carbon monoxide.

Effective bulk gettering by means of zirconium is hence best achieved by the use of two filaments, one at 400°C and other at 800°C.

An example of a coating getter is *Ceto*, made by sintering a mixture of a powdered alloy called Ceral with thorium powder at 900°C *in vacuo*. Ceral is a mixture of Cermischmetal (80% cerium, 20% lanthanum) and aluminium with the chemical composition (Ce, La)Al_2. The total getter composition is 20 Th, 5 Ce, 1 La, and 12 Al, in atoms (van Vucht [128]). The sintered bar of Ceto is ground to powder and made into a paste, with amyl acetate or nitrocellulose as a binder. This paste is painted onto a suitable part or electrode (often the anode or the inside of a metal envelope) of an electron tube with a coating thickness of 15 to 25 mg per sq cm. Activation of this getter is undertaken at 1,000°C, attained usually by induction heating the part, and the Ceto then sorbs and retains gas in large quantities in the temperature range 200 to 700°C. The capacity in torr-litre of 1 mg of Ceto at 20°C for various gases is 46×10^{-3} for hydrogen, 21×10^{-3} for oxygen, 3×10^{-3} for nitrogen, and 2×10^{-3} for carbon dioxide (Ehrke and Slack [129]).

Glasses. A great variety of glasses exists but only a few of them are generally encountered in vacuum work.

The simplest glass is 60 to 75% silicon dioxide (SiO_2), in the form of quartz sand, with 5 to 15% alkali oxide and 5 to 15% alkali-earth oxide. The commonest acidic oxides used in making other glasses are silicon dioxide and boric oxide (B_2O_3). Basic oxides include: lime (CaO); potash (K_2O); soda-lime (Na_2O); magnesia (MgO); and oxides of barium, zinc, manganese, lead, aluminium, iron, arsenic, lithium, tin, and zirconium.

Window glass, plate glass, and bottle glass are mixtures of silica, soda, and lime, to which alumina is sometimes added to improve the durability of the glass and diminish the risk of devitrification during manufacture.

Glasses most often encountered in vacuum technology are soda-lime glass, borosilicate glass, lead glass, and special glasses for sealing to metals. The constitutions of these glasses will depend somewhat on the manufacturer, but typical cases are:

Soda-lime: 75% SiO_2; 16% Na_2O; 7% CaO; 1% Al_2O_3; 1% Sb_2O_3.

The borosilicate glass, Pyrex: 80·2% SiO_2; 3·9% Na_2O; 0·3% K_2O; 0·2% CaO; 12% B_2O_3; 3% Al_2O_3; 0·3% As_2O_3.

Lead glass: 60% SiO_2; 8% Na_2O; 5% K_2O; 26% PbO; 0·1% Al_2O_3; 0·5% Sb_2O_3.

In making glass, the mixture of glass constituents is heated in a furnace to 1,400 to 1,500°C until it is free of bubbles, and is then poured into moulds or drawn into tubing or sheet glass. The subsequent cooling of the glass must be carefully regulated; an annealing schedule being followed depending on the nature and mass of the glass. Such annealing must always be done after the glass has been in the molten state, e.g. after glass-blowing. The hard borosilicate glasses such as Pyrex and Hysil can be worked in an oxygen-gas flame (about 1,400°C) in the form of thin wall tubing without careful subsequent annealing. Whereas the glass-blower usually works soda-lime glass in an air-gas flame (about 900°C), and subsequently cools it very carefully in a luminous gas flame without air, finally depositing a protective carbon deposit on the cooling glass from the coal-gas flame. If a complicated piece of glass apparatus, or pressed or moulded glass as in a valve base, is made then it must be annealed in an oven. Annealing schedules depend, firstly, on the hardness of the glass, which decides the high temperature given in Table 5.7, and,

TABLE 5.7

Annealing temperatures for glasses

Glass	*High annealing temperature in °C*	*Linear thermal expansion coefficient per degC*	*Density gram per cu cm*
Soda-lime	520	85 to 90×10^{-7}	2·5
Lead	430	86 to 94×10^{-7}	3·0
Pyrex	590	33×10^{-7}	2·24
Tungsten-sealing	580	40×10^{-7}	2·32
Kovar-sealing	480	48×10^{-7}	2·25
Molybdenum-sealing	600	45×10^{-7}	2·32

secondly, on the thickness and complexity of the glass component made, which determines the rate of cooling. The component should be heated in an oven so that the whole of it is uniformly at the high temperature (Table 5.7) for a short time. It is then cooled through about 100°C at a rate not exceeding $(20/d^2)$ degC per min for soft glasses (soda-lime and lead), and $(100/d^2)$ degC per min for hard glasses (borosilicates), where d is the wall thickness in mm. The rate of cooling down to room temperature after this 100°C drop can then be rapid, as no strain is introduced at this stage. It is best to cool comparatively slowly, however, as a complex shape may crack, even though no permanent strain is introduced.

If two glasses of different coefficients of linear thermal expansion are fused together, the joint will crack on cooling. A *graded seal* is essential to ensure a permanent seal with a number of intermediate glasses depending on the difference between the thermal expansion coefficients of the two outside glasses. Fusing a borosilicate glass such as Pyrex or Hysil to a soda-lime glass requires four intermediate glasses whose expansions increase progressively from that of Pyrex (the smallest) to that of soda-lime (the largest). These coefficients of thermal expansion between directly-joined glasses must be close together in value over a wide temperature range.

Glass-to-metal seals depend on the necessity that the fused metal must 'wet' the glass surface, and that the linear thermal coefficients of expansion are matched over the working range.

The strain in glass or in the glass of a glass-to-metal seal resulting from imperfect annealing is vividly demonstrated by a strain viewer. A polarized light beam from a sheet of polaroid is passed through the glass component and viewed by an eye-piece containing a Nicol prism or a second sheet of polaroid. A quarter-wave plate is also inserted in this optical system near the eye-piece. Strain in the glass then gives a coloured fringe effect in the field of view, clearly showing up the region of strain.

For making glass-to-metal seals with the soft soda-lime and lead glasses (linear thermal expansion coefficient: 9×10^{-6} per degC approximately at 20°C), the following metals may be used:

(*i*) platinum (expansion coefficient: $8{\cdot}9 \times 10^{-6}$ at 0°C and 11×10^{-6} at 800°C);

(*ii*) Dumet alloy or 'red platinum', a 43% Ni, 57% Fe alloy sheathed with copper and usually borated;

(*iii*) feathered copper edges as in the Housekeeper joint, where one end of a copper tube is flared out into a cone tapering to a feather edge of 0·001 to 0·002 inch thickness (the flared end of the copper tube is heated to form cuprous oxide which wets the glass tube fitted and fused to it);

(*iv*) an alloy of 50% Ni, 50% Fe, which seals to lead glasses A.E.I. C12, G.E.C. L1, and Jena 16 III, in wire diameters up to 5 mm;

(*v*) an alloy of 26% Cr, 74% Fe, which seals to Chance lead glass GW1 and soda glass GW2, Corning G5, G6, and G8, and G.E.C. L14;

(*vi*) an alloy of 42% Ni, 52% Fe, and 6% Cr, which seals to A.E.I. C12 and G.E.C. lead glass L1;

(*vii*) Fernichrome (37% Fe, 30% Ni, 25% Co, 8% Cr), which seals to Chance lead glass GW2 and Corning G5 and G8.

With hard glasses such as Pyrex (linear thermal expansion coefficient: $3{\cdot}3 \times 10^{-6}$ per degC at 20°C) and Hysil, direct sealing to tungsten ($4{\cdot}5 \times 10^{-6}$ at 0°C rising to 6×10^{-6} at 100°C) and molybdenum ($5{\cdot}5 \times 10^{-6}$ at 0°C) can be practised with wires up to 2 mm in diameter. However, it is preferable, especially for larger diameters, to use a tungsten-sealing borosilicate glass such as A.E.I. C14, Chance GS1 (Intasil), Corning 371 BN, G.E.C. W1, or Jena 1646 III; or for molybdenum a molybdenum sealing glass such as A.E.I. C14, Corning 704, Chance GS4, or Jena 1639 III.

Much use is made in vacuum technology of *Kovar*-to-glass seals. Kovar has the composition 53·7% Fe, 29% Ni, 17% Mo, and 0·3% Mn, and a melting point of 1,450°C. After annealing, its linear coefficient of thermal expansion is 5·7 to $6{\cdot}2 \times 10^{-6}$ per degC in the range 30 to 500°C. Nilo K, Fernico 1, Darwin F, and Vacon are all glass-sealing alloys similar to Kovar, not identical in composition but the same as regards glass-sealing properties.

Before sealing to glass, the Kovar should be degreased, rinsed in alcohol, dried, and heated to 900°C in a hydrogen furnace to clean and anneal it. If the Kovar is to be soldered to a second metal (so acting as an intermediary between this metal and glass), it is best copper-plated where the soldered joint is to be made and pure silver or silver-copper eutectic used, with fusion preferably in an hydrogen furnace. Kovar should never be heated to above 1,100°C.

Kovar is sealed to a special borosilicate glass such as Kodial, A.E.I. C40, or Chance GS3, having typically the composition: 67% SiO_2; 3% Na_2O; 4·6% K_2O; 21% B_2O_3; 4% Al_2O_3; 0·4% Sb_2O_3; and a thermal expansion coefficient of $4{\cdot}8 \times 10^{-6}$ per degC at 20°C. Such glasses can be sealed to the commoner borosilicate glasses by a graded glass seal. For example, Chance GS3 is sealed to Hysil or Pyrex via Chance GS4 and GS1 (Intasil) as intermediaries.

A metal can be soldered to glass (or porcelain or other ceramic) by metallizing the glass with platinum (p. 252) and soldering to the platinum. Better practice when joining a borosilicate-glass tube to a corresponding metal tube is first to fire-polish the end of the tube, coat it with silver oxide in a binder, and fire at 600 to 700°C to produce a dry silver deposit (Martin [130]). This silver layer is then burnished and soldered to the metal tube, which fits over the glass and has a wall thickness not exceeding 0·1 mm to permit differential expansion. The flux used is a moderately concentrated solution of tartaric acid in water to which is added one-third of its volume of glycerine. This flux is spread on the work and soldering is undertaken with a soldering-iron or by careful use of a gas-air flame.

Porcelain will seal directly to Pyrex glass on fusion in an oxygen-gas flame provided tube diameters do not exceed 0·5 inch. To seal silica or quartz to Pyrex glass, silver chloride fused at 450°C is employed.

If a fused joint between metal and glass is not possible, a greased or waxed, ground cone or flange joint may be used. Hard wax, such as W-wax, will readily effect a join between a glass tube inserted inside a metal tube of slightly greater diameter. Araldite may also be used provided the expansion coefficients of the glass and metal are not too dissimilar.

The viscosity, thermal expansion, and electrical resistivity of glass are composition sensitive, but the mechanical strength and thermal conductivity do not vary greatly from glass to glass.

Average breaking stress: 3 to 7 kg per sq mm.
Stress for 1% risk of breakage: 1 to 2 kg per sq mm.
Specific heat: 0·18 to 0·2 cal per gram per degC.
Thermal conductivity: 1·7 to 3×10^{-3} cal per cm per sec per degC.
Electrical resistivity: 10^{8} to 10^{16} ohm cm.

These figures are quoted for normal temperatures. Rise of temperature has little effect on tensile strength or specific heat, but the thermal and electrical conductivity both rise rapidly with temperature.

The viscosity of glass varies widely with temperature. At the melting or founding temperature, the viscosity is between 1,300 and 1,400 poise for soda-lime glass, but this figure varies widely with glass composition. At annealing temperatures (Table 5.7), which can be defined as at specific viscosities, values of 10^{13} to 10^{14} poise prevail.

Glass fluoresces under the action of ultra-violet, X-radiation, electron and ion bombardment, the fluorescence being very marked in devitrified glass, i.e. one in which small crystals form if the cooling schedule is wrong, so the viscosity is too high at certain temperatures.

The transmission of visible light through glass varies slightly with the type. Generally speaking, ultra-violet radiation at wavelengths below 3,000 Å is virtually cut off by a glass block 1 cm thick, though, in the case of some high-silica optical glasses, the transmission at 3,000 Å may be as much as 50%.

To clean glass thoroughly, the usual practice is to soak it in a solution consisting of 35 ml of saturated potassium bichromate ($K_2Cr_2O_7$), or preferably chromium trioxide (CrO_3), in 1 litre of concentrated sulphuric acid. The acid is poured slowly into the chromate or oxide solution, which is then used at 110°C. The chromate solution should be reddish in colour; if it becomes muddy or greenish, it is best discarded. Chromium trioxide is preferred because bichromate may result in alkaline salts remaining in crevices. The glass must be finally washed in distilled water (preferably warm) and dried in an oven or in a warm dust-free air stream.

When glass is melted *in vacuo* at 1,400°C, water vapour, oxygen, and carbon dioxide are evolved.

The quantity of gas desorbed from glass when it is heated *in vacuo* (e.g. during bake-out of a glass chamber undergoing pumping) is a function of the glass composition and the condition of storage since manufacture. Increase of alkali content increases the gas to be removed. Storage is best in a warm, dry place free of dust. Washing in soap and water alleviates the deleterious effects of storage but rarely eliminates it if a pattern is produced by weathering.

Glass sorbs gases and vapours. Physical adsorption of gases inert to the glass surface is with small binding energies, so such gas is readily removed by bake-out. However, the presence of alkali ions in the glass means that chemisorption of active gases is considerable, giving significantly large binding energies. For example, hydrogen ions, H^+, in the gas or vapour may replace sodium ions, Na^+, in the surface of a soda-lime glass. The initial heat of chemisorption of water at a clean glass surface is 10kcal per gram-molecule or more, resulting in coverage of the surface by OH groups. As this coverage increases, chemisorption gives place to physical adsorption with a lower binding energy of about 6 kcal per gram-molecule. Clean glass retains a few monolayers of active gas with considerable binding energies: additional gas is considerably more weakly bound to the gas covered surface. Holland [131] gives a full account of gas sorption by glass.

After glass has been cleaned chemically and dried, it is an advantage to bake it in air, because this drives off much gas and water vapour, so reducing considerably gas desorption on subsequent bake-out during pumping.

Sherwood [132] showed that the gases chiefly evolved by glass are water vapour and carbon dioxide. He determined the total gas evolution in cu mm at s.t.p. from a glass surface of area 350 sq cm (Table 5.8).

TABLE 5.8

Evolution of gas from glass baked *in vacuo*

Glass	*Cu mm at s.t.p. of gas evolved by a glass surface of area 350 sq cm at temperatures in °C noted*						
	100°	150°	200°	300°	400°	500°	600°
Soda-lime	0	36	22	8	6	11	50
Lead	6	10	10	7	8	12	30
Borosilicate	4	7	10	20	16	12	30

Note that the gas evolution reaches a maximum at 150° for soda-lime glass, 175°C for lead glass, and 300°C for borosilicate glass, and then decreases, subsequently to rise to a still higher value in all cases as the softening point of the glass is reached or approached.

At the lower temperatures up to 200 to 300°C, the physically adsorbed gas is evolved; as the temperature is raised to the softening point, almost all the gas is water vapour, arising from chemical decomposition of the glass.

Alpert [133] determined that the molecular gases, from borosilicate glass (Pyrex) unbaked on an ultra-high vacuum system, amount to 10^{-8} torr-litre per sec per sq cm, which decreased to 3×10^{-18} torr-litre per sec per sq cm after long term bake-out at 450°C, the gas evolved then being almost entirely water vapour. Further, due to helium permeation through the Pyrex from the atmosphere (p. 231), an influx of 10^{-15} torr-litre per sec per sq cm was measured, 3,000 times that due to water vapour.

Aluminosilicate glasses have much lower permeability for helium than the borosilicate glasses such as Pyrex and Hysil, and hence have been used in constructing vacuum systems to attain pressures below 10^{-11} torr. They have also the advantage that bake-out at 600°C or more can be practised. However, glass-blowing of aluminosilicate glasses is difficult and so is the manufacture of suitable metal-to-glass seals. For the latter purpose, molybdenum coated with chromium subsequently oxidized in wet hydrogen at 1,100°C has been sealed to aluminosilicate glass.

Gold. The chief present use of gold is as an O-ring metal gasket in bakeable ultra-high vacuum systems (section 3.8). It has also been used to make a diffusion seal between copper and other metals, e.g. in the construction of cavity magnetrons. The clean, accurately-machined, metal surfaces to be joined are either gold plated or separated by a thin gold foil (24 carat), pressed together and baked at a temperature well below the melting point of gold (1,063°C).

Gold has a very low secondary electron emission coefficient. Furthermore, it is inert and gas molecules have a very low sticking coefficient to gold. Electrodes in some electron tubes (e.g. the omegatron) are sometimes gold plated, because the absence of surface corrosion and low adherence of gas films ensures that no electrically polarized layer exists between the metal and the space outside (the vacuum). Consequently, potentials impressed on the electrodes are truly the potentials in the electric field immediately adjacent to the gold surface.

Gold can be coated on glass and other substrates by either vacuum deposition (p. 262) or sputtering (p. 255). A film of gold 5×10^{-6} cm thick on glass has selective transmission to green light. The transmission is 4·5% at 4,000 Å, 10% at 5,000 Å, 4% at 6,000 Å, and 2% at 7,000 Å.

A gold film, 60 to 70 Å thick, sputtered onto a reactively sputtered film of bismuth oxide of thickness 100 Å on glass provides, on heating to 200°C, a resistance of 10 ohm per square and an optical transmission of 70% for light from a tungsten lamp at 2,700°K. This provision and developments from it have led to the production of glass windows (e.g. for aircraft), which can be heated for demisting and defrosting by the passage of electric current through the gold film.

A colloidal suspension of gold in an organic liquid combined with an adhesion promoting agent (e.g. Johnson Matthey liquid bright gold 12% C and 12% Y) can be applied by brushing, dipping, or spraying to glass (using C) or ceramic (using Y), to form a film. This is baked in air to 520 to 650°C in the case of glass, or 650 to 850°C for a ceramic, so that the organic material is driven off to leave a bright, strongly-adherent, gold film.

Indium. This metal is valuable for making vacuum seals in cases where the gas and vapour evolution from elastomers is undesirable, and yet high bake-out temperatures are not demanded. Indium has a melting point of 156°C, a boiling point of 2,100°C, and a low vapour pressure in the molten state (10^{-5} torr at 670°C). Indium sealing gaskets are described in section 3.8. Bake-out up to 120°C is possible using these gaskets.

Indium can often be conveniently used as a low temperature solder which adheres well to the common metals. Indium-sealed vacuum valves or cut-offs have also been constructed where valve closure is obtained by a metal cap, which is lowered into the indium melted by a surrounding heater carrying electric current. On cooling, the indium solidifies round the cap.

Induction Heating. If a coil carrying a high frequency alternating current is placed near a conductor, eddy currents induced in the conductor will raise its temperature. If the conductor is within an evacuated glass chamber rather than in the atmosphere, the temperature attained will be much greater for a given power dissipation in the conductor, as heat loss is now mainly by radiation only. This fact is of great use in raising to high temperatures, for degassing purposes, the metallic or carbon structures in evacuated electron tubes. It has

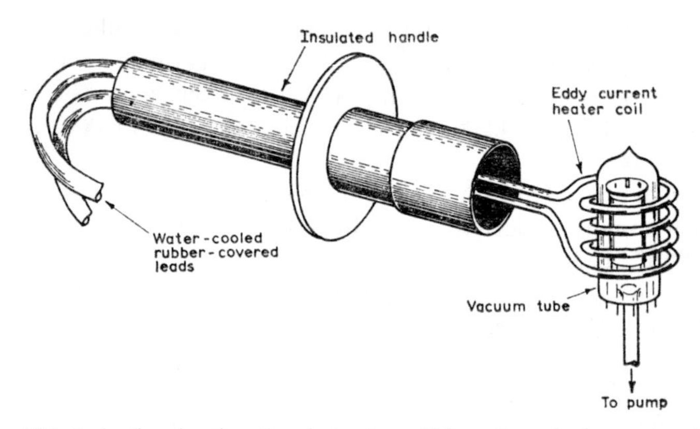

Fig. 106. Induction heating the electrodes within a thermionic vacuum tube undergoing pumping.

also been developed for heating and melting metals in vacuum metallurgy and for soldering and brazing techniques both *in vacuo* and in air.

The electrodes in a vacuum tube may be induction heated to 800 to 1,000°C or much greater temperatures (but excessive vaporization of the metal must be avoided), and gases which were not evolved during oven bake-out at, say, 450°C will be rapidly released (Fig. 106).

The eddy-currents induced in the metal will be subject to skin effect, whereby the depth of penetration for an alternating current will decrease the higher the frequency. For metal and carbon electrodes, a frequency of 200 to 500 kc per sec provided by a valve oscillator is usual, with a coil length not greater than four times its diameter.

An equation given by Reche [134] enables approximate calculations to be made for induction heating a hollow metal cylinder. If N

is the number of turns of the induction heater coil, which is placed around the evacuated tube containing the cylinder, and I is the r.m.s. current of frequency f, which is passed through this coil,

$$NI = \sqrt{\left(\frac{2\times 10^{-6}\, T^4 A}{G f^{1/2} \rho^{1/2}}\right)} \tag{5.3}$$

where T°K is the absolute temperature to which the cylinder is raised, A sq cm is surface area of this cylinder, and ρ is the resistivity in ohm-cm of the metal of the cylinder at the temperature T°K. G is a factor depending on D, the diameter of the circular induction heater coil, d, the diameter of the metal cylinder, and h, the height of this cylinder. Values of G are given in the Table 5.9.

TABLE 5.9

Evaluation of G in equation [5.3]

D/d	d/h	G
1·2	1·0	1·6
	0·5	0·9
	0·2	0·4
1·8	1·0	0·9
	0·5	0·7
	0·2	0·3

As an example, consider a cylinder of nickel of height 5 cm and diameter 2·5 cm, and so of surface area 39 sq cm, to be heated to 800°C, i.e. 1,073°K. At 1,073°K, the resistivity of nickel is 45×10^{-6} ohm cm.

Applying equation (5.3),

$$NI = \sqrt{\left(\frac{2\times 10^{-6}\times 1073^4\times 39}{45^{1/2}\times 10^{-3} G f^{1/2}}\right)}$$

Choosing a frequency f of 500 kc per sec, which will cause the majority of the heat to be dissipated in a thickness of metal of about 0·04 inch because of skin effect, and an induction heater coil of diameter 4·5 cm, it follows that $D/d=1{\cdot}8$, whereas $d/h=2{\cdot}5/5=0{\cdot}5$, so G from Table 5.9 is 0·7, and

$$NI = \sqrt{\left(\frac{78\times 10^{-6}\times 1{,}073^4}{6{\cdot}71\times 10^{-3}\times 0{\cdot}7\times (5\times 10^5)^{1/2}}\right)} = 5{,}580$$

Hence, a coil of 10 turns would require a current of 558 amp r.m.s.

at 500 kc per sec. A 2 kW induction heater is the usual size for degassing the electrodes in radio valves and cathode-ray tubes.

Mercury. Extensively used in manometers, McLeod gauges, mercury vapour pumps, cut-offs (Fig. 107*a*), switches (Fig. 107*b*), discharge lamps, etc.

Mercury readily combines with several metals at room temperature to form amalgams. Iron, steels, cobalt, nickel, platinum, rhodium, molybdenum, and tungsten are not readily affected.

Mercury can be cleaned by the following methods.

(*a*) Much dirt and grease can be removed from very dirty mercury by bubbling air (e.g. by the use of a water-jet pump) through it to form a readily removed scum on the surface.

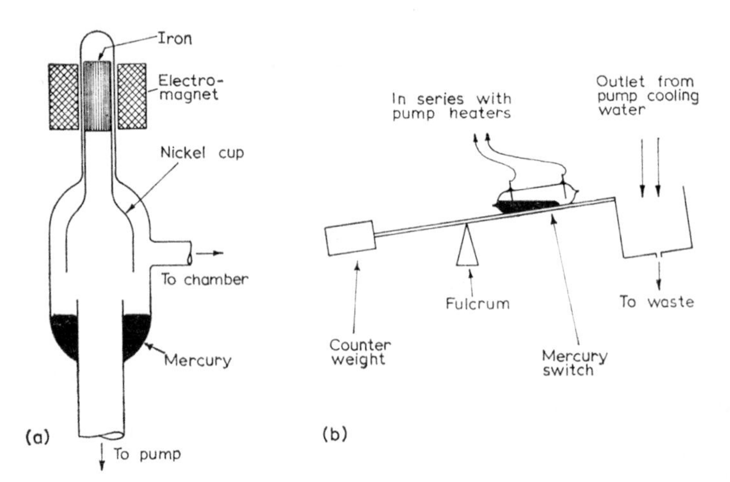

Fig. 107. (*a*) A mercury cut-off. (*b*) A mercury switch used for protecting against failure of cooling water supply to a vapour pump.

(*b*) Remove particles of foreign matter by squeezing through chamois leather.

(*c*) Remove metallic impurities by sprinkling down a long vertical column (glass tube 3 to 4 foot in height and about 1 inch in diameter) of nitric acid (3 parts water to 1 of concentrated acid by volume). This is preferably followed by sprinkling through a similar column of distilled water. The sprinkler is conveniently a glass funnel with a drawn-down outlet tube.

(*d*) Distil mercury *in vacuo*: the mercury is boiled in an electrically-heated borosilicate-glass container leading to a water-cooled Liebig condenser inclined at about 30° to the horizontal, with its outlet joined by a vacuum-tight seal to a collecting vessel. The interior of this apparatus is evacuated to about 10^{-2} torr by means of an oil-sealed mechanical rotary pump.

The last process should always be employed to clean mercury intended for use in pumps or gauges; indeed, double distillation is recommended. If the mercury is initially very dirty, stages (*a*) to (*d*) should be undertaken in sequence.

Mercury vapour can cause serious poisoning of the blood-stream if inhaled.

Density = 13·6 gram per cu cm at 20°C; melting point = −39°C; boiling point = 357°C at 760 torr; surface tension = 475 dyne per cm at 20°C *in vacuo*, and 400 to 500 in air.

Mica. An insulator, much used for supporting electrodes in vacuum tubes, as a dielectric in capacitors, and for general electrical insulation.

Mica occurs naturally in the forms of muscovite and biotite. Muscovite mica ($H_2KAl_3(SiO_4)_3$), which is practically colourless, can be obtained. Biotite mica is much discoloured.

Mica can be split along its cleavage planes into very thin sheets; thicknesses of 0·0005 inch are obtainable.

Before use in a vacuum tube (e.g. for electrode spacers) mica should be carefully cut to shape, rinsed in acetone, then alcohol, and finally baked in air to 200°C to remove water vapour. Mica surfaces can be regarded as saturated with gas, and mica contains about 18% of combined water.

If baked above 650°C, mica loses its normal form, the cleavage planes separate and a silver-grey appearance is assumed.

Electrical leakage across mica is often a source of difficulty; the cleaning and baking before use *in vacuo* reduces this leakage greatly. Careful handling after cleaning and baking is desirable, finger-prints being particularly deleterious.

Electrical leakage across mica, due to any volatilized films produced during heating of electrodes in a vacuum tube, is reduced by spraying alumina on the mica before assembly of the electrodes, so that any deposited metal is much less likely to form a continuous conducting surface.

Metal films evaporated onto mica do not adhere strongly; rubbing with a cloth will readily remove them.

A mica sheet, 0·001 inch thick, transmits approximately 90% of normally incident white light. Ultra-violet radiation at wavelengths below 2,600 Å is cut off.

To seal mica to metal or glass, e.g. to provide a vacuum-tight thin window on the end of a cylindrical glass or metal tube which is transparent to low energy beta-particles (end-window Geiger-Müller counter tube), other nuclear particles, X-rays, and ultra-violet radiation, a useful method (Donal [135]) is to use lead borosilicate glass of low flow temperature as a cement. The mica window cannot be comfortably more than 3 cm in diameter. It is made slightly larger than the squared-off and ground end of the glass tube over which it is placed. The low melting point glass is ground fine, mixed into a paste with water and painted on the tube end. The mica is pressed onto this preparation and the paste spread also around the edges of the mica to prevent them from splitting. Heating for some 10 min at 600°C and subsequent slow cooling makes a satisfactorily leak-free joint. The glass to which the mica is to be sealed should be selected to have a linear thermal expansion coefficient near to that of mica. The mica used was 0·005 to 0·02 inch in thickness.

Molybdenum. A refractory metal used particularly for transmitting valve anodes, X-ray tube targets, and grid wires in thermionic vacuum tubes. Wires down to 0·025 mm diameter can be drawn.

Its melting point of 2,625°C, combined with low vapour pressure at elevated temperatures (vapour pressure $=10^{-5}$ torr, at 1,987°C), and the fact that it is more ductile than tungsten makes it a suitable metal for heaters, heat shields, and supports. It will spot-weld to most metals but not to tungsten or to itself, though a thin grease film does enable a weak spot-weld to be made. It is degassed by vacuum – or hydrogen – stoving. It oxidizes very readily on heating in air. To clean it preparatory to making a metal-to-glass seal (it seals directly to Pyrex, but preferably to a molybdenum sealing glass, p. 239), it is heated to about 800°C in contact with sodium nitrite.

An alloy of 51% molybdenum with tungsten is ductile, shaped more easily than tungsten, and is stronger than molybdenum; it is used for heater wires especially in indirectly-heated thermionic cathodes. Molybdenum reacts chemically with oxygen which it sorbs readily at 1,000°C. When heated, it also chemisorbs hydrogen, C_2H_2, and C_2H_4, and sorbs nitrogen and carbon monoxide.

Movement of mechanism *in vacuo.* Three methods of imparting motion to a device in a vacuum from a mechanism outside in the atmosphere are shown in Fig. 108; their operation is self-explanatory. A means of inserting a small specimen into a vacuum without introducing air (Fig. 109*a*) is a vacuum-lock technique that can be extended to accommodate larger objects with attention to elastomer seal design. Two other convenient types of seal which can be used to enable a translatory or rotary motion to be imparted to a device *in vacuo* are the Wilson [136] seal (Fig. 109*b*) and the Gaco seal (Fig. 109*c*). The former uses a neoprene washer containing a central hole which has a diameter approximately two-thirds of that of the smooth cylindrical shaft which passes through it. This washer is flexed at the centre by the 30° conical bevel on the metal plate shown. Shaft

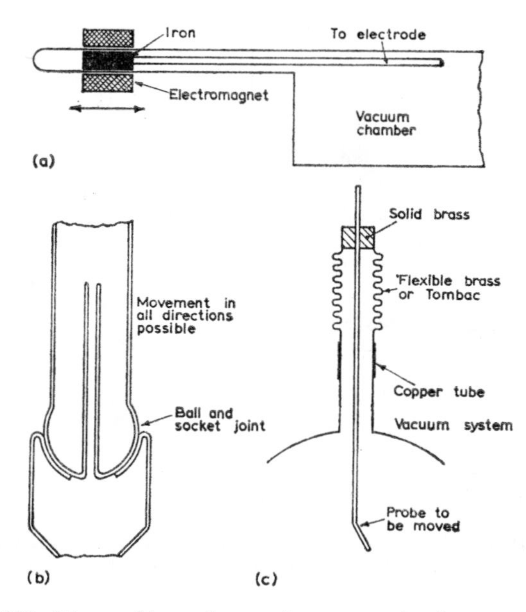

Fig. 108. Means of imparting motion to a mechanism *in vacuo*.

diameters can be from $\frac{1}{16}$ to 1 inch; light lubrication by vacuum grease is advantageous. A double Wilson seal, i.e. with two bevelled washers separated by a gap, is preferred if freedom from leakage is a problem, with a shaft subject to frequent movement. The space between the two washers can either be filled with a low vapour pressure silicone oil or can be evacuated by a rotary pump connected to a side-tube leading to this space. The single Wilson seal is, however, usually satisfactory at pressures down to 10^{-5} torr.

The Gaco seal is preferred by many workers, especially when continuous shaft rotation is practised. This is a specially-moulded neoprene gasket, which is spring loaded to ensure a tight seal against the shaft.

Nickel. Much used before World War II for the construction of vacuum-tube electrodes, but is expensive for this purpose and has now been replaced by stainless steel and other alloys. It has a fairly high melting point (1,455°C), is easily cleaned and degassed, resists corrosion, spot-welds readily to most metals and itself, and is very ductile when annealed. Often electroplated onto other metals, e.g. copper and mild steel, to render them more corrosion resistant for

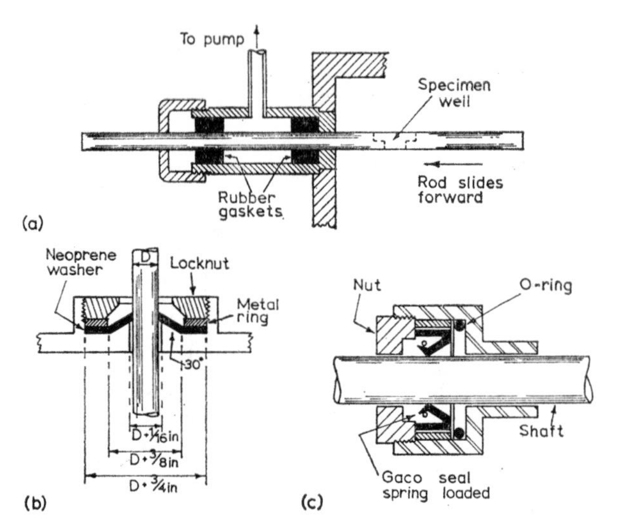

Fig. 109. (*a*) Inserting a specimen into vacuum through a vacuum lock. (*b*) The Wilson seal. (*c*) The Gaco seal.

high vacuum use. Can also be electroplated onto stainless steels, molybdenum, and tungsten. To nickel-plate iron and steel, copper is often deposited first.

Nickel is ferromagnetic with a maximum permeability of 700 gauss per oersted and a Curie point of 360°C.

Nichrome is a nickel alloy of composition 64% Ni, 11% Cr, 25% Fe, with a melting point of 1,350°C, which is much used for making electric heaters in ovens and furnaces (resistivity at 20°C $= 110 \times 10^{-6}$ ohm cm; temperature coefficient of resistance $= 1{\cdot}7 \times 10^{-4}$ per degC).

Chromel is an alloy of nickel and chromium with some iron, useful in making thermocouples with alumel. *Alumel* is a nickel alloy containing small amounts of iron and manganese. The Chromel-Alumel thermocouple (both metals can be hard soldered or spot-welded; both have a melting point of 1,400°C) can be used over the temperature range from −200 to 1,300°C. The e.m.f. established with a cold junction at 0°C and the hot junction at 100°C is 4·1 mV.

A group of nickel alloys with melting points usually between 1,300 and 1,350°C known as *Monel* contain variously 60 to 70% Ni, 30 to 35% Cu, and smaller additions. Some of these are hard and of high mechanical strength. K-Monel is non-magnetic.

Palladium. A noble metal of the platinum group. Chiefly used as a getter and a source of supply of hydrogen because palladium dissolves hydrogen to form palladium hydride: at s.t.p., 6 mg of hydrogen are sorbed by 100 gram of palladium. At −190°C, the sorption is greatly increased. If heated to above 300°C, the hydrogen is re-evolved. Also used as means of metering small amounts of hydrogen into a vacuum (p. 232), as gas permeability to hydrogen is significant and temperature dependent.

For gas sorption, the metal is used in the form of palladium black, prepared by dissolving in aqua regia, then baking, and adding sodium carbonate to free the compound of acid. The solution is then warmed, adding acetic acid and afterwards a warm concentrated solution of sodium formate. The palladium will precipitate in the form of palladium black, which has to be washed and dried. It oxidizes if heated in air and is soluble in nitric acid.

A palladium paste, A 286 (Johnson and Matthey Ltd.), can be brushed or sprayed onto unglazed ceramics then heated to 700 to 1,500°C to leave a palladium film. For coating glass, liquid palladium 4½% is used and baked at 560 to 650°C, whilst for glazed ceramics, liquid palladium 3349, baked at 650 to 800°C. Palladium begins to oxidize at 400°C, so an inert or reducing atmosphere is desirable during the final stages of baking.

Platinum. A noble metal, inert to all acids except aqua regia. Employed whenever long use without corrosion is desirable, and, like gold, is sometimes used for constructing electrodes where absence of surface contamination and gas layers avoids potential barriers from being formed, so that the potential in the vacuum immediately outside the electrode surface is the same – and remains the same – as that on the electrode itself. Seals readily to some common glasses to give vacuum-tight join. Often sandwiched between copper and glass to make a seal. Sputters readily but is difficult to evaporate *in vacuo*. Above 700°C, platinum foil shows considerable permeability to hydrogen and deuterium. It is very ductile and wires of diameters as small as 0·02 mm are obtainable. Can be used at its melting point (1,774°C) for brazing molybdenum and tungsten, and also as foil sandwiched between molybdenum or tungsten wires or ribbons to enable them to be spot-welded. Easily spot-welded.

A range of suspensions of platinum in organic liquids is available, which can be brushed or sprayed to form a film on glass or ceramic which is baked to leave a very strongly adherent platinum film. For glass, the best suspension is Johnson and Matthey Ltd. liquid platinum F104 which can be applied by brushing, dipping, or spraying and is baked at 560 to 650°C. For glazed ceramics, liquid platinum 7% is used and baked at 650 to 800°C; for unglazed ceramics, platinum paste N758 is applied by brushing and baked at 700 to 1,500°C.

Rhenium. The greater availability in recent years of this ductile metal in ribbon and thin foil form has led to its use as a thermionic emitter filament in place of tungsten, especially in hot-cathode ionization gauges and in the ion sources of mass spectrometers. Unlike tungsten, it is little affected by heating at atmospheric pressure and further resists corrosion and interaction with gases and vapours at elevated temperatures. It has excellent mechanical and electrical properties at high temperatures and spot-welds readily to most metals. The melting point is 3,180°C and its vapour pressure is 10^{-5} torr at 2,367°C. It is available in the forms of pure rhenium sheet, wire, and powder; rhenium/molybdenum alloy sheet, wire, and seamless tubing; rhenium/tungsten alloy sheet and wire; and also thoriated rhenium/tungsten wire.

Rhodium. A metal very similar to platinum, but cannot be readily drawn into wire form. Chief use is in electroplating, forming an excellent corrosion-free surface. A rhodium film vacuum-deposited by evaporation onto glass gives a very strongly adherent, hard film, which is almost neutral as an optical front-surface reflector and with an overall white-light (tungsten lamp) reflectivity of about 80%. Rhodium front-surface mirrors have been used for naval and aircraft optical instruments where severe weathering is likely. A semi-transparent thin film of rhodium on glass transmits almost neutrally in the region 4,000 to 8,000 Å.

Rubbers and Elastomers. Thick-walled, natural rubber tubing is often used for making temporary connections at pressures above 10^{-2} torr; it should not be used at lower pressures as it gives off hydrogen, has an effective vapour pressure of about 10^{-4} torr and exhibits some gas permeability. A low sulphur content is essential. Before use, it must be thoroughly clean: boiling in caustic soda solution followed by washing in distilled water and drying in a dry, oil-free, air blast is satisfactory.

In almost all vacuum practice, particularly for making O-rings, gaskets, and other seals for use at high vacuum, artificial rubbers or elastometers are used (section 3.7). Among the variety available, those most frequently encountered are the nitrile rubbers (Perbunan and Buna N), the chloroprene material neoprene, and the fluorocarbon Viton.

Nitrile rubbers are generally preferred to neoprene; both are oil-resistant but the former produces excellent quality moulded O-rings and other shapes, has lower gas permeability to nitrogen, and better resistance to compression than most elastomers. Heating of both nitrile rubbers and neoprene to about 120°C is possible *in vacuo*. On heating, water is predominantly desorbed initially, and hydrocarbons.

Viton A is resistant to high temperatures and has lower gas permeability than nitriles and neoprene, especially to hydrogen. It can be heated *in vacuo* to 200 to 250°C; water vapour, carbon monoxide, and carbon dioxide are desorbed. The low gas permeability and ability to withstand bake-out at up to 250°C (above 300°C decomposition products are desorbed, though evidence exists that this occurs at considerably lower temperatures) have lead to the use of Viton A as an O-ring and gasket material for vacuum systems operating at pressures down to 10^{-9} torr. However, it has poor resistance to compression, particularly at 250°C, so frequent replacement of the gasket is necessary if bake-out is practised.

Barton and Govier [137] have used a mass spectrometer to study the gases evolved by elastomers on degassing; data is also given by Bailey [115]. Turnbull, Barton, and Rivière [116] give data on gas permeability from which Table 5.10 is extracted.

TABLE 5.10

Gas permeability of some elastomers

Elastomer	*Gas permeability**				
	H_2	He	N_2	O_2	A
Nitrile rubber PB60	100	100	<1	8	17
Neoprene†	100 to 140	100	2·5 to 9·0	18·5 to 30	16 to 28
Natural rubber 337	450	290	70	147	200
Viton A	26·6	100	<1	<1	<1

* In 10^{-11} cu cm at 760 torr and 25°C times thickness in cm per sq cm (area of cross-section) per cm Hg pressure differential.

† Range of values is for different neoprenes: C.S. 2368B and C.S. 2367 were used.

Apart from the use of Viton A subjected to bake-out on the vacuum system, elastomer gaskets are in general not used in regions of systems where a pressure below 10^{-6} torr is to be obtained. They are hence usually confined to those parts of a vacuum system on the high pressure side of the cold trap.

Crawley and Csernatony [138] give information on the degassing characteristics of nitrile and Viton A O-rings of $1\frac{1}{4}$ inch nominal diameter, 0·139 inch cross-sectional diameter, and 60° shore hardness. Unbaked nitrile was shown to have about twice the degassing rate of Viton A, but the real advantages of Viton A showed up after a 4 hour bake of the test chamber at 100°C, when its degassing rate was about one-fiftieth of that of nitrile. Further, Viton A baked for 16 hour at 100°C gave a degassing rate of 3 to 4×10^{-10} torr-litre per sec per sq cm of surface, about the same as unbaked copper. However, baking Viton A at a higher temperature of 200°C causes higher degassing, probably because the evolution of water vapour then becomes complete, but other volatile constituents of the Viton A are evolved rapidly.

Viton A is of great interest as a sealing material for demountable joints in ultra-high vacuum systems where moderate bake-out temperatures of about 100°C are maintained; however, care with selection of this material is essential because samples from various manufacturers differ appreciably in properties.

Silica. Amorphous silicon dioxide; the crystalline variety is quartz, a mineral. Fused or vitreous silica is used for some transmitting valve envelopes, photocell tubes sensitive to ultra-violet, and in vacuum practice where a transparent insulating envelope is needed which can withstand excessive thermal shock because of its very low linear thermal expansion coefficient of $0{\cdot}42 \times 10^{-6}$ per degC up to 1,100°C. Silica spacers are also useful between electrodes and other parts in the assembly of electron guns, ion sources, etc. It is inert to all liquids except hydrofluoric acid.

Silica is harder than glass and most metals. Melting point is 1,700°C. It has to be worked in an oxygen-gas or oxygen-acetylene flame at temperatures at which it is necessary to protect the eyes of the glass-blower against radiation with dark glasses. Recrystallization tends to occur at 1,200°C. Lead is the only metal which fuses satisfactorily to silica.

Quartz is much used in optics when high ultra-violet transmission is required, useful transmission over the wavelength range from 2,000 to 35,000 Å being obtained. A thin layer of silica evaporated onto optically-worked glass *in vacuo* can be used to protect the glass against the chemical action of weathering, but silicon monoxide – a useful vacuum deposited dielectric film – is much easier to evaporate (p. 262).

Silica exhibits higher permeability for helium than glasses (p. 229), and also for hydrogen, neon, and argon.

Silver. Readily sputtered and evaporated *in vacuo* to form a metallic film on a substrate which is not strongly adherent if the substrate is glass, but where an optically-polished surface so coated forms the best front-surface reflector for white light, from the point of view of reflection coefficient. Percentage of normally incident light reflected is 30% at 3,000 Å, 86% at 4,000 Å, and 94% in the range 6,000 to 12,000 Å.

Silver can be readily deposited onto glass by the chemical methods known as the Rochelle salt and the Brashear processes.

It is the best conductor of heat and electricity, making it very difficult to spot-weld electrically to itself or other metals. However, can be readily soldered and brazed. Silver is a useful brazing material in the pure form; there is also a range of hard silver solders among which silver-copper eutectic alloy (Ag 72%, Cu 28%) is available, especially for soldering copper.

Silver has been used in place of copper or gold for making bakeable metal gaskets for use in ultra-high vacuum.

Sputtering. When a positive ion is incident upon a metal surface, it may well eject an atom of the metal from this surface, a phenomenon known as sputtering. The sputtering coefficient, defined as the number of metal atoms ejected per incident ion, will depend on the nature of the incident ion (usually of a gas such as nitrogen, argon, hydrogen, etc.), its energy, the nature of the metal, and the angle of incidence of the ion, where the coefficient increases with increasing angle, i.e. is larger for oblique incidence. Sputtering coefficients range from vanishingly small values (sputtering negligible) up to about 10.

Sputtering is important in gaseous discharge phenomena and plays a part in the mechanism of sorption of ionized gas at metallic surfaces. It is utilized in the sputter-ion or Penning cold-cathode getter-ion pump (section 1.14) and has also been developed as a means of preparing thin films on substrates.

Fig. 110(*a*) shows a laboratory plant for sputtering a metal and depositing a film on a substrate, e.g. of glass; Fig. 110(*b*) is a unit for reactive sputtering. In the former, often known as *cathodic sputtering*,

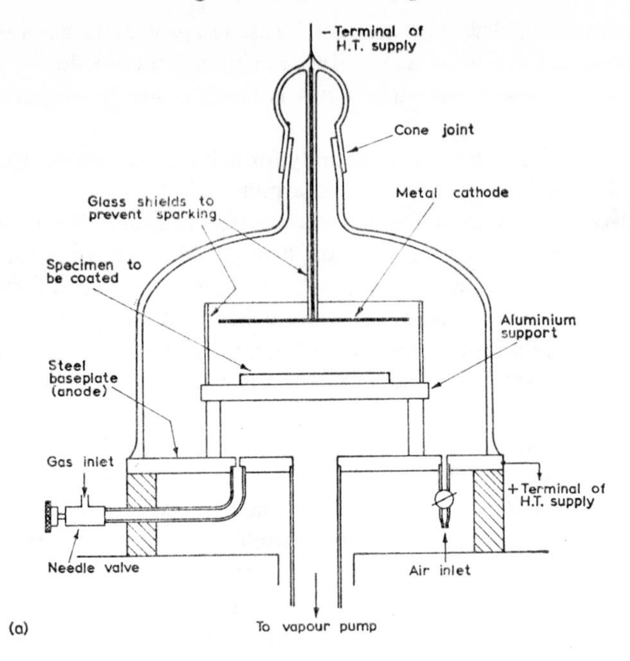

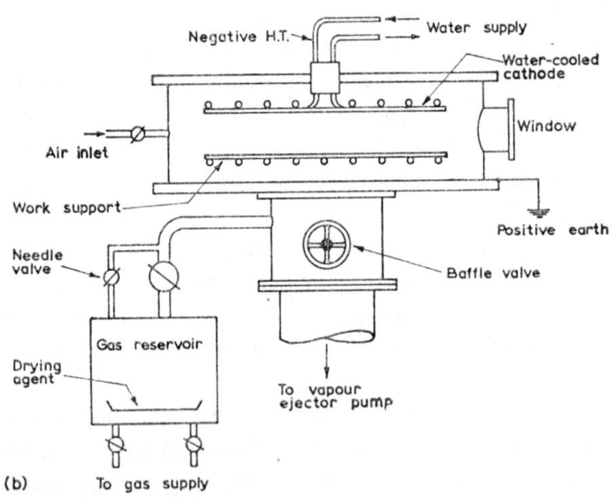

Fig. 110. (*a*) A laboratory plant for sputtering. (*b*) A unit for reactive sputtering.

the cathode is usually a plane sheet of the metal to be sputtered, or may be of copper or steel heavily electroplated with the metal. The gas pressure is adjusted to be between 10^{-2} and 10^{-1} torr. To achieve this pressure, it is best to pump the chamber first to about 10^{-5} torr by means of a rotary/vapour pump combination, and admit the gas (preferably an inert gas such as argon) through a needle valve to obtain an equilibrium between the inlet throughput and the throughput to the pumps. Across the cathode and anode (which is the steel base plate of the chamber) is maintained a steady p.d. of 1 to 2 kV with the anode earthed. Positive ions created in the discharge impinge on the cathode and eject from it atoms of the cathode metal, which then travel to the substrate to be coated. It is important to provide a H.T. power-pack for the anode-cathode voltage which has a high internal impedance (a resistor of 2 to 3 kΩ is often included), to ensure that the discharge does not become unstable and develop into an arc.

The sputtering rate is proportional to the current density: with a p.d. of 2 kV a current density of 0·5 to 1·0 mA per sq cm of cathode area is satisfactory. The coating uniformity is improved the larger the ratio of cathode diameter to cathode-substrate separation; this latter separation is so that the substrate is just outside the cathode dark space, which has a width of about 2 to 3 cm at the pressures used.

The sputtering coefficient increases with the energy of the incident positive ions (and hence with the anode-cathode p.d.) and with their mass. Though all metals can be sputtered, some, like aluminium and magnesium, offer difficulty because of surface oxide films. For obtaining metallic films, the best gas to use is generally argon, and those metals which are often more conveniently sputtered rather than evaporated *in vacuo* are gold, platinum, and palladium. Silver has a high sputtering coefficient but is usually evaporated to make a film. The sputtering rates of metals in descending order in an argon atmosphere are Cd, Ag, Pb, Au, Sb, Sn, Bi, Cu, Pt, Ni, Fe, W, Zn, Si, Al, Mg.

Reactive sputtering (Fig. 110*b*) is a technique for obtaining films of metal oxides or other metallic salts on a substrate. To form oxide films of a metal, the cathode is made of this metal and the glow discharge is maintained in an atmosphere consisting of a mixture of an inert gas (e.g. argon) and about 5% oxygen, at a total pressure of 10^{-1} torr. The gas mixture is best prepared in an auxiliary reservoir containing a drying agent (usually phosphorus pentoxide), and the sputtering atmosphere needs to remain uncontaminated, so that high speed pumping (usually by a vapour ejector pump) is needed with a relatively high rate of gas inlet to reduce the influence of gases desorbed from the chamber walls. The cathode-anode p.d. is generally about 3 kV with a discharge current density of 5 mA per sq cm.

A wide range of metals can be sputtered reactively to produce on a glass substrate a strongly adherent film of the corresponding oxide. Considerable work has been done on oxide films of bismuth, cadmium, copper, iron, tin, indium, and titanium; other metals which have been reactively sputtered are Al, Be, Ce, Pb, Mg, Mo, Ni, Pt, Si, Ta, Te, Th, W, and Zn. By controlling the partial pressure of oxygen, the degree of oxidation of the deposit can be decided; indeed, oxide films with excess metal atoms can be made, a technique of interest in semiconductor physics.

Steels. Mild steels consisting of iron mixed with a small percentage of carbon are much employed in the construction of large vacuum systems. These low carbon steels are readily machined, spot-welded, soldered, brazed, and welded. There are, however, many alloys of steel with very diverse properties. Though mild steel is adequate for the construction of vacuum chambers to be pumped down to 10^{-6} torr, provided it has a low sulphur content, which is necessary to minimize outgassing, the oxides and hydroxides which form on the surface sorb water which it is difficult to remove, particularly as satisfactory bake-out under vacuum is impractical. It is good practice, therefore, to plate the surface of the steel, which is exposed to vacuum, with nickel or chromium. For the many classifications of steels and basic applications reference is best to 'Wrought Steels' (*British Standards Institution*, 970 : 1955).

Stainless steels have become widely used for electrodes, tubing, including flexible corrugated tubing, vacuum valves, and chambers, especially in ultra-high vacuum technique. The austenitic ('Staybrite') 18/8 stainless steels, which are non-magnetic, of welding quality (i.e. can be welded to give good quality leak-free joints by means of argon-arc or heliarc techniques) and have fair-to-good free machining (i.e. can be machined without special cutting tools), are preferred. The 18/8 stainless steels contain approximately 18% chromium and 8% nickel (hence the name) with about 0·1% carbon and the remainder iron. Those containing small amounts of the additional metals such as molybdenum, titanium, and columbium are particularly resistant to corrosion. There are several varieties of 18/8 stainless steels: for vacuum work, those often chosen are En58B (melting point: 1430°C), which is titanium stabilized, En58F (melting point: 1,450°C), and En58J for tubing (melting point: 1,430°C). These stainless steels do not require heat treatment after welding.

These 18/8 stainless steels are more difficult to machine and braze than low carbon steels, and electric argon-arc welding, vacuum brazing, or electron-beam welding are desirable if to be used for constructing ultra-high vacuum systems and components. However, they have the great advantages that they resist corrosion, provide a smooth surface (readily electropolished) free of water-sorbing oxides and hydroxides, and can withstand bake-out for degassing up to 450°C or even to 700°C.

Unbaked stainless steel will evolve molecular gases (chiefly hydrogen, water vapour, and carbon monoxide) to an extent variously estimated to be between 10^{-9} and 10^{-7} torr-litre per sec per sq cm of surface exposed to ultra-high vacuum; the lower figure of 10^{-9} being obtainable with a very clean electropolished surface after some hours of pumping. After bake-out at 400°C for some 16 hour, electropolished stainless steel in an assembled ultra-high vacuum system will evolve primarily carbon monoxide and hydrogen, at a rate of about 5×10^{-12} torr-litre per sec per sq cm, which can be decreased by a factor of about 100 with a very prolonged bake-out. Consider an electropolished stainless steel vacuum chamber with an interior surface area of 1,000 sq cm. If unbaked, the evolution of gas from this surface will be at least 10^{-6} torr-litre per sec, demanding a pumping speed of $10^{-6}/10^{-9}$, i.e. 1,000 litre per sec to maintain a pressure of 10^{-9} torr. If baked, however, and assuming no other source of outgassing or leakage is present, the pumping speed requirement after bake-out is reduced to roughly $5 \times 10^{-9}/10^{-9}$, i.e. 5 litre per sec, and the maintenance of very low pressures becomes practicable.

Tantalum. A highly refractory metal with a melting point of 2,850°C. It is ductile, malleable, and has similar machinability (carbon tetrachloride cooling is used) to mild steel. Tantalum offers great resistance to corrosion and chemical attack, withstanding all acids except hydrofluoric and fuming sulphuric at normal temperatures, but is attacked by strong alkalis. The mechanical strength is comparable with that of steel, both in the fine wire and thin sheet forms. Electric spot-welding of tantalum to itself and a number of other metals is readily possible, and it has been used in the form of 0·001 inch thick foil as an intermediary in spot-welding tungsten and molybdenum.

Used for making transmitter valve anodes and other vacuum tube electrodes, especially where a large heat dissipation occurs. Vacuum-stoving must be practised and not hydrogen-stoving, because it sorbs hydrogen into solution strongly at elevated temperatures to become very brittle.

Pirani [139] found that tantalum sorbs about 740 times its own volume of hydrogen at s.t.p. when heated to 800°C; later work has established that hydrogen sorption increases in rate with temperature in the range 500 to 1,200°C, with an optimum take-up at 600°C. Oxygen is sorbed at 750°C, and rapidly at 1,500°C, up to a maximum amount equal to 20 times the metal volume at s.t.p.; above 1,500°C, the oxide Ta_2O is formed. Nitrogen is sorbed slowly below 1,300°C, but at 1,750°C the volume can be 80 times the metal volume at s.t.p. Tantalum has been used as a bulk getter: the powder is sintered on a thin strip of ribbon welded to a molybdenum or tungsten support which is heated *in vacuo*.

Thermionic Cathodes. A pure metal heated *in vacuo* to a temperature T°K will give an emission I in amp per sq cm decided by the Richardson-Dushman equation

$$I = AT^2\exp(-w/kT) \qquad (5.4)$$

where w is the work function energy of the metal in electron-volt (eV), and k is Boltzmann's constant (k=1/11,600 eV per degK). A is expressed in amp per sq cm per deg^{2}K and is a constant for a given metal but varies somewhat from one metal to another.

To obtain adequate electron emission, either w must be small or T large. The pure refractory metals used most frequently as high temperature emitters are tungsten (melting point 3,380°C; vapour pressure$=10^{-5}$ torr at 2,547°C) and rhenium (melting point 3,180°C; vapour pressure$=10^{-5}$ torr at 2,367°C). For the polycrystalline metal, the work function of tungsten is 4·52 eV, and that of rhenium is 5·1 eV. At 1,630°C, the emission from tungsten is $2{\cdot}3\times10^{-1}$ mA per sq cm, which rises to 298 mA per sq cm at 2,230°C; for rhenium the values are 10^{-1} mA per sq cm at 1,950°C, rising to 1 mA per sq cm at 2,150°C. Though tungsten is widely used, rhenium is recommended for ionization gauges and ion sources in mass spectrometers, because it exhibits much less reaction with gas at low pressures, and is free of the formation of carbides. Further it is not seriously oxidized, like tungsten, on exposure, when hot, to air at pressures exceeding 10^{-3} torr; it has excellent ductility, and does not become brittle on heating *in vacuo*.

Thoriated tungsten filaments and also thoriated rhenium filaments are both used, again with a preference for the latter. Thoriated tungsten filaments are prepared from tungsten wire containing 1 to 2% thoria (ThO_2). The activation process necessary for the filament within the evacuated tube is: (*a*) heat to 2,500°C for about 1 min to reduce the thorium oxide to thorium; and (*b*) reduce the temperature to about 2,000°C and maintain for some min to cause migration of the thorium to the surface. At an operating temperature of 1,730°C, the emission is 2·75 amp per sq cm, the work function being 2·6 eV. Thoriated rhenium is similarly activated but at lower temperatures, and gives about the same emission at 1,600°C.

A 50% thoria/50% rhenium combination is a useful emitter which, at 1,360°C, gives about 275 times greater emission per sq cm than thoriated tungsten (Esperson [140]). At 1,400°C, the emission is 100 mA per sq cm.

Oxide-coated cathodes consist of a mixture of about 60% barium carbonate and 40% barium carbonate sprayed onto a filament of tungsten, rhenium, or nickel alloy, or on a cathode of nickel alloy or rhenium, containing a heater of tungsten-molybdenum or rhenium-molybdenum alloy. The carbonate mixture is initially in colloidal suspension in alcohol, with a binder such as nitrocellulose. The spraying is by a compressed air gun or, alternatively, coating is by dipping, painting, or a cataphoretic process.

A typical schedule for such an oxide-coated cathode, of about 8 watt (heater voltage × heater current) consumption within a small electron tube on a vacuum system, is: (*a*) bake-out the tube in an oven at 350 to 450°C for 10 min; (*b*) induction heat the electrodes to 800°C for 20 to 90 sec; (*c*) switch on the cathode heater and raise its temperature to 1,150°C approximately for 10 to 100 sec, to drive off carbon dioxide from the carbonates, to leave strontium and barium oxides; (*d*) induction heat the electrodes whilst the cathode is still hot, to ensure that gas is not interchanged between the cathode and electrodes; and (*e*) seal-off the tube from the pumps and fire the barium-aluminium getter.

After seal-off, the cathode is activated by running it at 900°C approximately and drawing a current of about 100 mA for 2 to 5 min to the electrodes connected together at a suitable positive potential. The normal electrode operating potentials are then maintained for about 30 min. This activation causes an electrolytic action whereby free barium, which is the active emitter, forms within a matrix of the oxides at the surface. An 8 watt cathode should give an emission of 500 mA. The work function of the activated oxide-coated cathode is 1 volt approximately.

Oxide-coated cathodes are spoilt if air is admitted to the tube; thoriated tungsten cathodes need to be reactivated. Two low temperature emitters which withstand fairly well such air admission are tungsten or rhenium, coated with thoria, and an emitter consisting of a coating of lanthanum hexaboride (LaB_6) on tungsten, or preferably iridium or rhenium. Lanthanum hexaboride emitters have a work function of 2·66 eV and provide an emission of 100 mA per sq cm at 1,250°C. Coated onto rhenium, a life of 1,000 hour is possible.

Low temperature emitters such as lanthanum-hexaboride-coated rhenium operating at 1,250°C give considerable advantages in ionization gauges and mass spectrometers compared with tungsten, normally run at 2,200°C to give the same emission. Apart from the fact that radiant heat is less, so that outgassing of nearby electrodes is not so troublesome (e.g. in a mass spectrometer ion source), molecules of gases such as hydrogen, water vapour, pump oil vapours, and hydrocarbons are very much less likely to dissociate at a surface at a temperature of 1,250°C than at one at 2,200°C.

Vacuum Deposition of Thin Films. A wide range of metals and non-metallic materials can be heated *in vacuo* so that they evaporate and deposit on a nearby substrate. There are a great number of applications of such vacuum coating techniques; for a full account see Holland [112]. A laboratory vacuum coating plant (Fig. 111*a*) suitable for vacuum deposition at a residual pressure of 10^{-5} torr makes use of a filament heater or evaporator boat. The filament heater (Fig. 111*b*) consists of three strands of 0·5 to 1 mm diameter tungsten wire, which is very convenient for the evaporation of aluminium. The boat (Fig. 111*c*) is made of molybdenum and is suitable for several metals, silicon monoxide, magnesium fluoride, and other non-metallic substances.

The preparation of a front-surface aluminium mirror illustrates the main features of the process. Before introduction into the vacuum chamber, the optically-polished glass is cleaned thoroughly. Among several possible cleaning schedules a rigorous one is: (*a*) wash the glass with detergent and water; (*b*) rinse with freshly distilled water; (*c*) soak for some time in 10% potassium bichromate solution; (*d*) rinse with distilled water again; (*e*) wash with pure alcohol; and (*f*) polish with fresh clean cotton-wool or a silk cloth.

Meanwhile, the aluminium evaporator is prepared. Small U-shaped pieces of 1 mm diameter pure (99·9%) aluminium wire are placed at intervals over the three-strand tungsten heater, or a length of this wire is coiled round the heater. A pressure of 10^{-5} torr or below is then established in the chamber, and the tungsten heater temperature is raised to about 1,100°C by the passage of a.c. directly through it from a step-down transformer with a Variac regulator in the primary. The molten aluminium has a high surface tension of about 600 dyne per cm and wets the tungsten freely. After cooling, air is admitted to the chamber, which is removed, and the clean, polished glass is mounted above the evaporator on a suitable support. A backing pressure of about 10^{-2} torr is then re-established in the chamber, and the glass is subjected to vigorous ion bombardment cleaning. To undertake this, a discharge is set up in the residual gas between two ring-shaped aluminium electrodes mounted on suitable base-plate terminals. For a 12 inch diameter chamber, a discharge

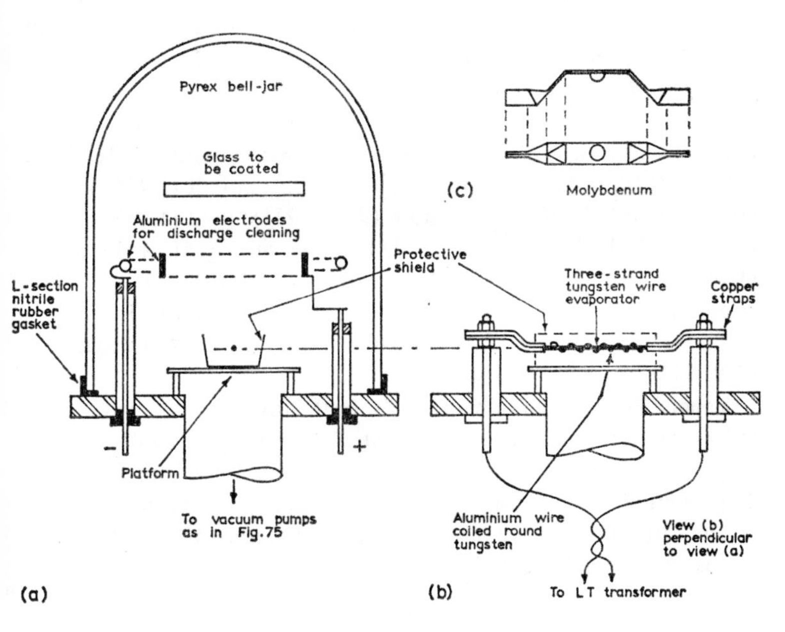

Fig. 111. Vacuum coating plant.

current of 100 mA at 3 kV is suitable. The H.T. power-pack used must have significant internal impedance (2,000 ohm, say) to prevent the discharge from becoming an unstable arc between surface prominences. The action of the energetic positive ions on the glass is somewhat complex, but the effect is to provide a surface to which aluminium films adhere strongly.

After ion bombardment cleaning for 10 to 20 min, the pressure is reduced to 10^{-5} torr or below by the vapour diffusion pump. The evaporator temperature is then raised quickly to 1,200 to 1,400°C, when the aluminium will evaporate rapidly to give an opaque film in 20 to 60 sec.

After admitting air and removing the coated glass, the film is left standing for some hour to permit the growth on its surface of a film of aluminium oxide (Al_2O_3). Such a film will be strongly adherent to the glass and withstand cleaning with cotton-wool or a Selvyt cloth.

Several other metals can be evaporated *in vacuo*, but most of them from a molybdenum boat rather than the tungsten-wire heater. An idea of the possibilities is gained by the following list of metals, against each of which is given, first, the melting point, and, second, the temperature in °C at which its vapour pressure is 10^{-2} torr, but generally higher temperatures than these are needed for a satisfactory evaporation rate.

Aluminium (660, 996); antimony (630, 678); barium (717, 629); beryllium (1,284, 1,246); bismuth (271, 698); cadmium (321, 264); calcium (810, 605); chromium (1,900, 1,205); cobalt (1,478, 1,649); copper (1,083, 1,273); germanium (959, 1,251); gold (1,063, 1,465); iron (1,535, 1,447); lead (328, 718); magnesium (651, 443); nickel (1,455, 1,510); platinum (1,774, 2,090); rhodium (1,967, 2,149); selenium (217, 234); silicon (1,410, 1,343); silver (961, 1,047); tin (232, 1,189); titanium (1,727, 1,546); uranium (1,132, 1,898); zinc (419, 343).

Vapour Pressures of Some Liquids. In the following list, the saturated vapour pressure is given in torr at 20°C. Acetone, 184·8; benzene, 74·65; carbon tetrachloride, 91·0; chloroform, 159·6; ethyl alcohol, 43·9; methyl alcohol, 96·0; ethyl ether, 422·2; ethyl bromide, 386·0.

The saturated vapour pressure in torr of water (or ice) at various temperatures is given in Table 5.11.

TABLE 5.11

Vapour pressure of water at various temperatures

Temperature in °C	100	80	60	40	20	0	−10	−20
Vapour pressure	760	355	150	55·3	17·5	4·6	1·95	0·77

Temperature in °C	−30	−40	−50	−60	−80
Vapour pressure	0·28	0·093	0·029	0·007	0·0004

REFERENCES

[1] GAEDE, W., *see* MEYER, G., *Verh. d. deutsch Phys. Ges.*, **10**, 753 (1907)

[2] GAEDE, W., *see* DUNKEL, M., (Ed.), *Wolfgang Gaede – Eine Schrift aus dem Nachlass* (E. Leybold, Cologne, 1951)

[3] POWER, B. D., and KENNA, R. A., *Vacuum*, **5**, 35 (1955, published 1957)

[4] GAEDE, W., *Ann. der Physik*, **46**, 357 (1915)

[5] LANGMUIR, I., *Phys. Rev.*, **8**, 48 (1916)

[6] ALEXANDER, P., *J. sci. Instrum.*, **23**, 11 (1946)

[7] FLORESCU, N. A., *Vakuum-Technik*, **12**, 255 (1963)

[8] JAECKEL, R., *Kleinste Drucke, ihre Messung und Erzeugung* (Springer, Berlin, 1950)

[9] JAECKEL, R., NÖLLER, H. G., and KUTSCHER, H., *Vakuum-Technik*, **3**, 1 (1954)

[10] NÖLLER, H. G., *Vacuum*, **5**, 59 (1955, published 1957)

[11] BURCH, C. R., *Proc. roy. Soc.*, **123**, 171 (1929)

[12] HICKMAN, K. C. D., *Rev. sci. Instrum.*, **1**, 140 (1930); and *J. Frank. Inst.*, **221**, 215 (1936)

[13] LATHAM, D., POWER, B. D., and DENNIS, N. T. M., *Vacuum*, **2**, 33 (1952, published 1953)

[14] HUNTRESS, A. R., SMITH, A. L., POWER, B. D., and DENNIS, N. T. M., *1957 Fourth Natl. Symp. Vac. Tech. Trans.*, p. 104 (Pergamon Press, Oxford, 1958)

[15] HICKMAN, K. C. D., *1961 Trans. Eighth Natl. Vac. Symp. and Second Internatl. Congr.*, Vol. 1, p. 307 (Pergamon Press, Oxford, 1962)

[16] HICKMAN, K. C. D., *J. appl. Phys.*, **11**, 305 (1940)

[17] CRAWFORD, W., *Phys. Rev.*, **10**, 557 (1917)

[18] POWER, B. D., and CRAWLEY, D. J., *Vacuum*, **4**, 415 (1954, published 1957)

[19] CRAWLEY, D. J., and MILLER, J. M., *Vacuum*, **15**, 183 (1965)

[20] BIONDI, M. A., *Rev. sci. Instrum.*, **30**, 831 (1959)

[21] VENEMA, A., and BANDRINGA, M., *Philips tech. Rev.*, **20**, 145 (1958–9)

[22] POST, R. F., *see* ALPERT, D., *Handbuch der Physik* (Flügge, S., Ed.) Vol. 12, *Thermodynamik der Gase* (Springer, Berlin, 1958)

[23] CLOSE, K. J., HODGES, E. B., and ATCHISON, F., *Vacuum*, **16**, 385 (1966)

[24] GAEDE, W., *Phys. Zeit.*, **13**, 864 (1912)

[25] HOLWECK, M., *Comptes Rendus*, **177**, 43 (1923)

[26] SIEGBAHN, M., *Arch. Math. Astr. Phys.* (Roy. Swedish Acad.), **30B**, No. 2 (1943)

[27] BECKER, W., *Vakuum-Technik*, **7**, 149 (1958)
[28] HOLLAND, L., *J. sci. Instrum.*, **36**, 105 (1959)
[29] GOULD, C. L., and DRYDEN, R. A., *1961 Trans. Eighth Natl. Vac. Symp. and Second Internatl. Congr.*, Vol. 1, p. 369 (Pergamon Press, Oxford, 1962)
[30] HALL, L. D., *Rev. sci. Instrum.*, **29**, 367 (1958)
[31] JEPSEN, R. L., *Le Vide*, **14**, 80 (1959)
[32] BRUBAKER, W. M., *1959 Trans. Sixth Natl. Symp. Vac. Tech.*, p. 302 (Pergamon Press, Oxford, 1959)
[33] JEPSEN, R. L., FRANCIS, A. B., RUTHERFORD, S. L., and KIETZMANN, B. E., *1960 Trans. Seventh Natl. Vac. Symp.*, p. 45 (Pergamon Press, Oxford, 1961)
[34] WEHNER, G. K., *J. appl. Phys.*, **30**, 1762 (1959)
[35] CLAUSING, R. E., *1961 Trans. Eighth Natl. Vac. Symp. and Second Internatl. Congr.*, Vol. 1, p. 345 (Pergamon Press, Oxford, 1962)
[36] HOLLAND, L., and HARTE, A., *Vacuum*, **10**, 133 (1960)
[37] BANNOCK, R. R., *Vacuum.*, **12**, 101 (1962)
[38] MAINWARING, E. E., *Vacuum*, **14**, 303 (1964)
[39] FORTH, H. J., *Vakuum-Technik*, **10**, 227 (1961)
[40] HONIG, R. E., *1961 Trans. Eighth Natl. Vac. Symp. and Second Internatl. Congr.*, Vol. 2, p. 1166 (Pergamon Press, Oxford, 1962)
[41] MCLEOD, H. G., *Phil. Mag.*, **48**, 110 (1874)
[42] GAEDE, W., *Ann. der Phys.*, **46**, 357 (1915)
[43] ISHII, H., and NAKAYAMA, K., *1961 Trans. Eighth Natl. Vac. Symp. and Second Internatl. Congr.*, Vol. 1, p. 519 (Pergamon Press, Oxford, 1962)
[44] MEINKE, C., and REICH, G., *Vakuum-Technik*, **12**, 79 (1963)
[45] LECK, J. H., *Pressure Measurement in Vacuum Systems*, p. 8, (Chapman and Hall, London, 1964)
[46] ROSENBERG, P., *Rev. sci. Instrum.*, **10**, 131 (1939)
[47] JANSEN, C. G. J., and VENEMA, A., *Vacuum*, **9**, 219 (1959)
[48] PODGURSKI, H. H., and DAVIS, F. N., *Vacuum*, **10**, 377 (1960)
[49] BARR, W. E., and ANHORN, V. J., *Instruments*, **19**, 666 (1946)
[50] FLOSDORF, E. W., *Ind. Eng. Chem.*, **10**, 534 (1938)
[51] PIRANI, M., *Verh. der Deutsch Phys. Ges.*, **8**, 686 (1906)
[52] MEYER, W., *ATM Z.*, **3**, 117 (1938)
[53] WEISE, E., *Z. techn. Phys.*, **18**, 73 (1937)
[54] BECKER, J. A., GREEN, C. B., and PEARSON, G. L., *Trans. Amer. Inst. Elect. Engrs.*, **65**, 711 (1946)
[55] GRUBER, H., *Vakuum-Technik*, **3**, 65 (1954)
[56] VARIĆAK, M., *1961 Trans. Eighth Natl. Vac. Symp. and Second Internatl. Congr.*, Vol. 1, p. 483 (Pergamon Press, Oxford, 1962)
[57] ROBERTS, R. W., MCELLIGOTT, P. E., and JERNAKOFF, G., *J. Vac. Sci. Tech.*, **1**, 62 (1964)
[58] VOEGE, W., *Phys. Z.*, **7**, 498 (1906)
[59] KNUDSEN, M., *Ann. der Phys.* (*Leipzig*), **32**, 809 (1910)
[60] LECK, J. H., *Pressure Measurement in Vacuum Systems*, p. 135 (Chapman and Hall, London, 1964)

[61] DUMOND., J. W. M., and PICKELS, W. M., *Rev. Sci. Instrum.*, **6**, 362 (1935)
[62] PENNING, F. M., *Philips techn. Rev.*, **2**, 201 (1937), and *Physica*, **4**, 71 (1937)
[63] PENNING, F. M., and NIENHUIS, K., *Philips techn. Rev.*, **11**, 116 (1949)
[64] REDHEAD, P. A., *Adv. Vac. Sci. Tech. Proc. First Internatl. Congr.*, Vol. 1, p. 410 (Pergamon Press, Oxford, 1960)
[65] BUCKLEY, O. E., *Proc. Natl. Acad. Sci. U.S.A.*, **2**, 683 (1916)
[66] DUSHMAN, S., and FOUND, C. G., *Phys. Rev.*, **17**, 7 (1921), and **23**, 734 (1924)
[67] MORSE, R. S., and BOWIE, R. M., *Rev. sci. Instrum.*, **11**, 91 (1940)
[68] NOTTINGHAM, W. B., *see* DUSHMAN, S., *Scientific Foundations of Vacuum Technique*, 2nd. edition (Lafferty, J. M., Ed.), p. 330 (Wiley, New York, 1962)
[69] BAYARD, R. T., and ALPERT, D., *Rev. sci. Instrum.*, **21**, 571 (1950)
[70] CARTER, G., and LECK, J. H., *Brit. J. appl. Phys.*, **10**, 364 (1959)
[71] BLEARS, J., *Proc. roy. Soc. A.*, **188**, 62 (1947)
[72] WEINREICH, O. A., and BLEECHER, H., *Rev. sci. Instrum.*, **23**, 56 (1952)
[73] STECKELMACHER, W., and VAN DER MEER, S., *J. sci. Instrum.*, **27**, 189 (1950)
[74] HOLMES, J. C., *Rev. sci. Instrum.*, **28**, 290 (1957)
[75] BENTON, H. B., *Rev. sci. Instrum.*, **30**, 887 (1959)
[76] CLOSE, K. J., and HODGES, E. B., *Vacuum* **16**, 75 (1966)
[77] ALLENDEN, D., *Electronic Engineering*, p. 31 (Jan. 1958)
[78] YARWOOD, J., and CLOSE, K. J., *Introductory Electricity and Atomic Physics* (Longmans, London, 1964)
[79] CLOSE, K. J., *Private communication*
[80] REDHEAD, P. A., *Rev. sci. Instrum.*, **31**, 343 (1960)
[81] SCHULZ, G. J., and PHELPS, A. V., *Rev. sci. Instrum.*, **28**, 1051 (1957)
[82] LAFFERTY, J. M., *J. appl. Phys.*, **32**, 424 (1961)
[83] LECK, J. H., *Pressure Measurement in Vacuum Systems*, p. 82 (Chapman and Hall, London, 1964)
[84] DOWNING, J. R., and MELLEN, G., *Rev. sci. Instrum.*, **17**, 218 (1946)
[85] VACCA, R. H., *1956 Vac. Symp. Trans. Amer. Vac. Soc.*, p. 93 (Pergamon Press, Oxford, 1957)
[86] ROEHRIG, J. R., and VANDERSCHMIDT, G. F., *1959 Vac. Symp. Trans. Amer. Vac. Soc.*, p. 82 (Pergamon Press, Oxford, 1960)
[87] BLANC, D., and DAGNAC, R., *Vacuum*, **14**, 145 (1964)
[88] GÜNTHER, K. G., *Vacuum*, **10**, 293 (1960)
[89] NIER, A. O., *Rev. sci. Instrum.*, **11**, 212 (1940), and **18**, 398 (1947)
[90] DEMPSTER, A. J., *Phys. Rev.*, **11**, 316 (1918)
[91] SOMMER, H., THOMAS, H. A., and HIPPLE, J. A., *Phys. Rev.*, **82**, 697 (1951)
[92] ALPERT, D., and BURITZ, R. S., *J. appl. Phys.*, **25**, 202 (1954)
[93] ALPERT, D., *J. appl. Phys.*, **24**, 860 (1953)

[94] ALPERT, D., *Rev. sci. Instrum.*, **22**, 536 (1951)
[95] BAKER, D., *Vacuum*, **12**, 99 (1962)
[96] HOLDEN, J., HOLLAND, L., and LAURENSON, L., *J. sci. Instrum.*, **36**, 281 (1959)
[97] BRYMNER, R., and STECKELMACHER, W., *J. sci. Instrum.*, **36**, 278 (1959)
[98] HICKAM, W. M., *Rev. sci. Instrum.*, **20**, 472 (1949)
[99] GROVE, D. J., *1958 Fifth Natl. Vac. Symp. Trans.*, p. 9 (Pergamon Press, Oxford, 1959)
[100] VENEMA, A., *Adv. Vac. Sci. Tech. Proc. First Internatl. Congr.*, Vol. 1, p. 389 (Pergamon Press, Oxford, 1960)
[101] POWER, B., DENNIS, N. T. M., and CSERNATONY, L. de, *1963 Trans. Tenth Natl. Vac. Symp.*, p. 147 (Macmillan, New York, 1963)
[102] RIVIÈRE, J. C., and ALLINSON, J. D., *Vacuum*, **14**, 97 (1964)
[103] LECK, J. H., *Pressure Measurement in Vacuum Systems*, p. 170 (Chapman and Hall, London, 1964)
[104] NORMAND, C. E., *1961 Trans. Eighth Natl. Vac. Symp. and Second Internatl. Congr.*, Vol. 1, p. 534 (Pergamon Press, Oxford, 1962)
[105] ROEHRIG, J. R., and SIMONS, J. C., *1961 Trans. Eighth Natl. Vac. Symp. and Second Internatl. Congr.*, Vol. 1, p. 511 (Pergamon Press, Oxford, 1962)
[106] FEAKES, F., and TORNEY, F. L., *1963 Trans. Tenth Natl. Vac. Symp.*, p. 257 (Macmillan, New York, 1963)
[107] NELSON, H., *Rev. sci. Instrum.*, **16**, 273 (1945)
[108] OCHERT, N., and STECKELMACHER, W., *Vacuum*, **2**, 125 (1952)
[109] WHITE, W. C., and HICKEY, J. S., *Electronics*, **21**, 100 (1948)
[110] POWER, B. D., and CRAWLEY, D. J., *Vacuum*, **4**, 415 (1954, published 1957)
[111] HASS, G., *J. opt. Soc. Amer.*, **39**, 532 (1949)
[112] HOLLAND, L., *Vacuum Deposition of Thin Films* (Chapman and Hall, London, 1956)
[113] BENNETT, A. I., *Vacuum*, **3**, 43 (1953, published 1954)
[114] DUSHMAN, S., *Scientific Foundations of Vacuum Technique*, 2nd edition (Lafferty, J. M., Ed.) (Wiley, New York, 1962)
[115] BAILEY, J. R., *Handbook of Vacuum Physics*, Vol. 3, Part 4 (Beck, A. H., Ed.) (Pergamon Press, Oxford, 1964)
[116] TURNBULL, A. H., BARTON, R. S., and RIVIÈRE, J. C., *An Introduction to Vacuum Technique* (Newnes, London, 1962)
[117] KOHL, W. H., *Materials and Techniques for Electron Tubes*, Chaps. 2 and 4 (Reinhold, New York, 1960)
[118] BONDLEY, R. J., *Electronics*, **20**, 97 (July, 1947)
[119] HOUSEKEEPER, H., *J. Amer. Inst. Elect. Engrs.*, **42**, 954 (1923)
[120] BOWER, J. H., *J. Res. Natl. Bur. Stand.*, **33**, 200 (1944)
[121] HONIG, R. E., and HOCK, H. O., *R.C.A. Review*, **21**, 360 (1960)
[122] NORTON, F. J., *1961 Trans. Eighth Natl. Vac. Symp. and Second Internatl. Congr.*, Vol. 1, p. 8 (Pergamon Press, Oxford, 1962)
[123] WALDSCHMIDT, E., *Metall*, **8**, 749 (1954)

[124] ESPE, W., *Werkstoffkunde der Hochvakuumtechnik*, Vol. I, p. 893 (VEB Deutscher Verlag, Berlin, 1959)
[125] ESCHBACH, H. L., *Praktikum der Hochvakuumtechnik* (Akad. Verlag. Geest and Portig K.–G, Leipzig, 1962)
[126] JOSSEM, E. L., *Rev. sci. Instrum.*, **11**, 164 (1940)
[127] PORTA, DELLA, P., *Vacuum*, **4**, 284 and 464 (1954, published 1957)
[128] VUCHT, VAN, J. H. N., *Vacuum*, **10**, 170 (1960)
[129] EHRKE, L. F., and SLACK, C. M., *J. appl. Phys.*, **11**, 129 (1940)
[130] MARTIN, S. T., *Rev. sci. Instrum.*, **11**, 205 (1940)
[131] HOLLAND, L., *The Properties of Glass Surfaces* (Chapman and Hall, London, 1964)
[132] SHERWOOD, R. G., *J. Amer. chem. Soc.*, **40**, 1645 (1918)
[133] ALPERT, D., *Vacuum*, **9**, 91 (1959)
[134] RECHE, K., *Wiss. Veröffent. Siemens-Konzern*, **12**, 1 (1933)
[135] DONAL, J. S., *Rev. sci. Instrum.*, **13**, 266 (1942)
[136] WILSON, R. R., *Rev. sci. Instrum.*, **12**, 91 (1941)
[137] BARTON, R. S., and GOVIER, R. P., *J. Vac. Sci. Tech.*, **2**, 113 (1965)
[138] CRAWLEY, D. J., and CSERNATONY, L. de, *Vacuum*, **14**, 7 (1964)
[139] PIRANI, M., *Z. Elektrochem.*, **11**, 555 (1905)
[140] ESPERSON, G. A., *J. appl. Phys.*, **21**, 261 (1950)

INDEX